Political Approaches
to Injury Control
at the
State Level

# Political Approaches to Injury Control at the State Level

Abraham B. Bergman, M.D., Editor

Sue Lockett John, Technical Editor
Corrine Condie, Production Supervisor

University of Washington Press

Seattle and London

Proceedings of a conference held on January 26, 1991,
at the Stouffer Madison Hotel, Seattle, Washington.

This conference was sponsored by the Harborview Injury Prevention and Research Center of the University of Washington, in association with the Trauma Foundation of San Francisco. The generous support of Allstate and Safeco Insurance companies to conduct the conference and publish the proceedings is acknowledged.

Copyright © 1992 by the University of Washington Press
Printed in the United States of America

All rights reserved. No part of this publication may be reproduced or transmitted in any form or by any means, electronic or mechanical, including photocopy, recording, or any information storage or retrieval system, without permission in writing from the publisher.

Library of Congress Cataloging-in-Publication Data
Political approaches to injury control at the state level / Abraham B. Bergman, editor.
    p. cm.
Proceedings of a conference held in Seattle, Washington, Jan. 26, 1991, sponsored by the Harborview Injury Prevention and Research Center of the University of Washington in association with the Trauma Foundation of San Francisco.
ISBN 0-295-97176-2 (alk. paper)
    1. Accidents—Prevention—Government policy—United States—States—Congresses. 2. Wounds and injuries—Prevention—Government policy—United States—States—Congresses. 3. Safety regulations—United States—States—Congresses. I. Bergman, Abraham B., 1932– . II. University of Washington. Harborview Injury Prevention and Research Center. III. Trauma Foundation.
HV677.U5P65    1992                                 91-40227
363.1'07'0973—dc20                                           CIP

The paper used in this publication meets the minimum requirements of American National Standard for Information Sciences—Permanence of Paper for Printed Library Materials, ANSI Z39.48–1984.

Contents

Introduction
Abraham Bergman................................................. vii

An Overview Of Legislative/Regulatory Interventions
Fred Rivara....................................................... 3

The Effectiveness Of Legal Countermeasures
Against Alcohol-Impaired Driving
Allan Williams.................................................... 17

DWI Legislation: The Minnesota Experience
Nancy Bode....................................................... 27

The Art (And Necessity) Of Coalition Building:
The California Alcohol Tax Initiative
Andrew McGuire................................................... 37

A Campaign To Curtail Irresponsible Alcohol
Advertising In Washington State
Abraham B. Bergman............................................... 47

The Politics of Compulsory Motorcycle Helmet
Legislation: The Successful Washington Experience
Roy Ferguson..................................................... 51

The Almost Successful California Experience: What
We And Others Can Learn From It
Elizabeth McLoughlin............................................. 57

Efforts To Prevent Burns From Hot Tap Water
Murray Katcher................................................... 69

The Case Of The Fire Safe Cigarette: The Synergism
Between State And Federal Legislation
Andrew McGuire................................................... 79

**Legislative Approach To Child Drowning Prevention**
Garen Wintemute.................................................. 89

**The 1988 Maryland Gun Law: An Example Of Public Health Advocacy**
Steven Teret...................................................... 99

**Concluding Remarks**
Abraham B. Bergman............................................. 109

**Contributors** 111

# Introduction

It is a truism that politicians have the capacity to save many more lives than physicians and other health professionals. No where is this more evident than at Harborview Medical Center, a major trauma center where I work. Here we are forced do deal with a never-ending deluge of bodies and spirits broken in large measure by products of our modern society such as cars, motorcycles, and guns. We have Ralph Nader and his colleagues to thank for an era in the 1960s and '70s where a host of laws were passed and regulations implemented, especially in the traffic safety field, that led to significant reductions in trauma mortality and morbidity.

Through friendship with the late Senator from Washington, Warren Magnuson, then Chairman of the Commerce Committee, I was privileged to have a front row seat for observing the practice of what I call "political medicine." For example, Magnuson's legislation creating the Consumer Product Safety Commission has resulted in the prevention of numerous product related injuries; his Flammable Fabrics Act Amendments (1967) markedly reduced the number of children sustaining flame burns from their clothing; and his Poison Prevention Packaging Act (1970) dramatically reduced the number of children hospitalized with ingestions.

With the ascension of President Reagan, and political action committees exercising increasing influence over Congress, the consumer protection movement in Washington, D.C. fell upon hard times in the 1980s. That situation persists today. As relatively more political power is moving to the states, more interest is being shown in how consumer protection and safety legislation can be achieved at the state level.

By bringing together a diverse group of experts in the the fields of injury control and professional politics, I hope that the proceedings of this conference will serve as a practical "how-to" manual for those interested in employing the political process to control injuries in various states. Our two main objectives are: 1) In terms of mortality and morbidity, to define the most significant injury problems amenable to approach through legislation, regulation, and or legal means at the state level, and 2) To develop practical implementation strategies and useful data that can be employed at the state level. Our hope is that these proceedings will be most useful to workers in the injury prevention and consumer protection fields, as well as to public officials.

ABRAHAM B. BERGMAN, M.D.

# Political Approaches to Injury Control at the State Level

# An Overview of Legislative/Regulatory Interventions

Fred Rivara, MD, MPH

My task is to try and outline the injury problems for which legislative/regulatory approaches might be appropriate and successful. I'll not recite the usual litany of injury statistics with which we are all so familiar. Instead, I will lay out what I think are possible interventions for the six leading causes of serious injuries and deaths: motor vehicles, fires/burns, drowning, poisoning, firearms and falls. I'll try to examine what might be the impact on mortality for some of these interventions.

The basic philosophy I took when I approached this task, and which I think most of us use, is that there is no magic bullet to injury prevention. Rather, reductions in the more than 140,000 deaths each year from trauma will occur from small reductions in many different areas.

## Motor Vehicles

When considering motor vehicle injuries, it's important to break the problem down into its component parts: occupants, motorcyclists, and pedestrians. Although some interventions cut across types, such as anti-drunk driving programs, many are specific to the particular injury problems.

### Occupants: Drivers

The use of regulation and legislation to decrease injuries is certainly nothing new when it comes to drivers and occupants of motor vehicles. There are many examples of interventions which have had a significant impact on injuries. What are some interventions which remain to be implemented nationally?

Teen drivers, particularly young drivers under the age of 17, have a much greater risk of motor vehicle crashes than do older drivers. Raising the minimum age to 17 will reduce motor vehicle fatalities in younger drivers by as much as 85% (Williams, Karpf and Zador, 1985), without significantly increasing the risk for drivers above this age, i.e. delaying the effects of inexperience. The net effect would be to save 1375 deaths and 160,000 injuries per year among teens.

Teen drivers also have a much higher rate of fatal nighttime crashes than do older drivers. Among 16-17 year-old drivers, 48% of male and 44% of female deaths on this age group occur between 9 p.m. and 6 a.m. Studies indicate that curfew laws limiting the nighttime driving of newly licensed drivers can reduce crashes and fatalities by as much as 80%. Thus, of the 2000 occupant deaths each year in this age group, up to 800 may be prevented through curfew laws. Moreover, surveys performed by the Insurance Institute for Highway Safety (IIHS) and others indicate teens in this age group are readily accepting of such curfews. It's embarrassing when your mother tells you to be home by midnight, but not when it is the law.

Leon Robertson has clearly shown driver's education does not make teenagers safer drivers (Robertson, 1980). In fact, it has a net harmful effect because it allows teens to be licensed a year earlier than they would otherwise be without driver's ed. (Robertson, 1983). Elimination of driver's eduction would result in a decrease of over 650 deaths per year and a savings of more than $800 million (Rice and MacKenzie, 1989).

Repeal of the Right Turn on Red (RTOR) laws which were introduced throughout the U.S. during the 1970s would appear to be an appropriate intervention to protect occupants and pedestrians. Zador and colleagues from the IIHS have estimated that RTOR laws result in a 20% increase in motor vehicle crashes involving a right turn at signalized intersections and a 57% increase in pedestrian - motor vehicle collisions at these sites (Zador, Moshman and Marcus, 1982). Pedestrian collisions in urban areas increased by 70%. An increase of over 30% was found for child pedestrian injuries, 100% for adults and 110% for elderly pedestrians after adoption of RTOR. Approximately 1200 pedestrians are killed at intersections; a reduction of this number by 700 might be expected with repeal of RTOR laws. Whatever the savings in gasoline and time from RTOR, they are associated with substantial increases in pedestrian and other motor vehicle collisions.

Legislation to require motor vehicle occupants to wear seat restraints has been passed in over one-half of the states and has been successful in increasing seat restraint use. Use in states with laws averages about 50%, ranging from 30 to 75% (NHTSA, 1989). This contrasts to usage rates of approximately 30% in areas without mandatory use laws. Such laws result in an approximate 10% reduction in fatalities (Chorba, Reinfurt and Hulka,

1988). In addition to extending legislation to states presently without such laws, there is a need to close the gaps in coverage in states with existing laws. Allan Williams and his colleagues from the IIHS recently found, for example, that 26 of the 31 jurisdictions covered as of December 1987 exempted all rear seat passengers other than children (Wells, Williams, Fields, 1989). Six states excluded light trucks and vans, vehicles which are over-represented in crashes; this accounts for about 500 deaths per year. Laws in many states are secondary laws, i.e. police officers may only issue citations for failure to use belts if the motorist is first stopped for some other offense. Laws that permit primary enforcement have a significantly greater impact on seat belt use and thus fatalities than those which permit only secondary enforcement (Wagenaar, Maybee and Sullivan, 1988). Attention should be given to changing the seat belt laws in these states.

Approximately 20% of the population of the U.S. is still not covered by seat belt laws. Extending similar coverage to the individuals will result in approximately a 10% decrease in fatalities in this group, or approximately 650 deaths prevented (31,000 x 0.2 x 0.1).

One of the more controversial motor vehicle laws in recent years has been the decision by Congress in 1987 to allow states to increase speeds on rural interstates from 55 mph to 65 mph. Since that time, more than 38 states have increased the speed limit on most sections of their rural interstate highways. In the first few months after the laws were changed, there was a 15% increase in rural interstate fatalities. Fatalities increased by 26% to 29% in the second year of 65 mph speed limits on rural interstate highways, accounting for more than 500 excess deaths per year (Baum, Wells and Lund, 1990). This is far higher than predicted prior to the legislative change and indicates that these laws should be repealed.

## Occupants: DWI

Approximately 24,000 or 50% of motor vehicle fatalities involve alcohol. Most of these drivers were not just a little drunk; in 1986, the average BAC of drinking drivers involved in fatal crashes was 0.15. At our trauma center, of those individuals admitted with trauma who have positive BACs, the average level is over 0.18. Fortunately, there have been significant decreases recently in the proportion of motor vehicle fatalities in which the driver was intoxicated (BAC $\geq$ .10): from 30% in 1982 to 26% in 1986, a 14% decrease. There was a 28% decrease in the proportion of teen drivers involved in fatalities who were intoxicated during the same period.

Allan Williams will speak more about the effectiveness of different DWI legislation. Effective legislative options include administrative revocation of licenses, roadside breath testing at sobriety checkpoints, mandatory jail sentences for DWI, and an increase in the alcohol excise tax. The effect of these laws is not well known. However, there appears to be reasonable evidence that administrative revocation laws (present in 23 states) decrease nighttime or alcohol-related crashes by 10-15% (Nichols and Ross, 1989). Zador et al., estimate that administrative revocations could overall reduce fatal crashes by 5%; applied to the 50% of the population unprotected, this could save an additional 1125 lives per year (Rice and MacKenzie, 1989). Compared to other sanctions, license actions have the greatest individual and general deterrent potential. Sobriety checkpoints may also become more common with the recent Supreme Court decision. One study of checkpoints in Virginia estimated that this reduced alcohol related crashes by about 15% (Voas et al., 1988). If this was true generally, it could lower fatalities by a further 3600 per year.

**Occupants: Vehicles**

Changes in vehicles have been successful in reducing motor vehicle injuries and fatalities (Robertson, 1983). However, further vehicle changes could bring about substantial additional reductions in severe and fatal injuries. Seat belts offer far less protection in side impact than frontal collisions (Evans, 1988); side impacts accounted for 32% of all motor vehicle occupant fatalities in 1985. NHTSA estimated that up to 1100 fatalities could be prevented with better side-impact protection (NHTSA, 1988), rules which are still pending. Another recent study by the IIHS seems to corroborate this with estimates that 42% of fatalities in side impact could be prevented with better car design.

Passive seat restraints are now required in all cars for the front seat occupants. Extensions of airbags to include front and rear seat passengers will extend protection particularly in frontal collisions for the 50% of the population who ride in motor vehicles unrestrained. Use of full front seat airbags alone would reduce deaths by an estimated 6190 per year (Rice and MacKenzie, 1989).

Approximately 10% of passenger car severe and fatal injuries occur to occupants of the rear seat (NHTSA, 1990), amounting to approximately 2300 people in 1985 (NHTSA, 1987). Until recently, the rear seat was fitted with only lap belts, which are much less effective than are lap shoulder harnessess in preventing fatal injury (18% compared to 43%)

(Evans and Frick, 1988). In addition to a higher risk of fatal injury, occupants using lap belts only are at a higher risk of sustaining Chance fractures to the lumbar spine and injuries to the intestine and colon than are patients restrained with lap/shoulder belts (Anderson et al, 1991). Although NHTSA has mandated that rear seats be fitted with lap/shoulder belts in the outboard position, children cannot use the shoulder harness with current designs and are at increased risk of these lap belt associated injuries. Mandating a redesign of the belts to accommodate children and small adults is called for.

## Motorcycles

The most important intervention for motorcycle safety is clearly mandatory helmet use laws. Laws requiring motorcyclists to use helmets reduce motorcycle deaths by about 24-30% (Rice and MacKenzie, 1989). These laws increase helmet use from approximately 50% to close to 100%. The number of motorcyclist fatalities in states without helmet laws in 1985 was 2714. Therefore, based on a 24% reduction, it is estimated that 650 fewer deaths and almost 2000 fewer serious head injuries would have occurred had these states had helmet use laws. The estimated savings from motorcycle helmet laws would be $393 million annually using the human capital approach. Our studies at Harborview Medical Center (HMC) indicate 63% of the costs for the care of motorcyclists at trauma centers are paid from public funds, making it an important societal issue (Rivara et al., 1988).

The other area of motorcycle safety which I think can be impacted by regulation is that of superbikes. These are actually racing bikes made street legal through minor modifications. Some are capable of accelerating from 0 to 60 mph in 3 seconds with top speeds of greater than 160 mph. A study by the IIHS showed that the death rate of these cycles is twice that of other cycles (IIHS, 1987). These bikes should simply be banned.

## Pedestrian Injuries

Approximately 8000 individuals die each year in the U.S. from pedestrian injuries, representing 20% of all motor vehicle fatalities and the second leading cause of unintentional trauma deaths behind motor vehicle occupant deaths. The extent to which regulation and legislation can play a significant role in pedestrian safety is a question which remains to be answered. Clearly, RTOR laws, discussed previously, increase the risk to the pedestrian. Repeal of these laws would result in a reduction of 700 deaths per year. The effectiveness of other laws to decrease risk of pedestrian fatalities is unknown. Legislation which has been suggested includes "edge striping" to define the edge of the roadway, and laws regulating the position of school bus stops. However, enforcement of existing laws appears to be as crucial as any intervention. In Pennsylvania, a change in the law to improve pedestrian safety resulted in virtually no change in driver behavior with regard to pedestrian motor vehicle conflicts because the laws went virtually unenforced (Haight and Olsen, 1981). Here in Seattle, the emphasis has traditionally been placed on enforcement of jaywalking laws with very little enforcement against the motorists (8000 compared to 600 citations each year). However, with a clarification of the laws regarding pedestrian right of way and cooperation of the police department, more than 350 motorists per month are currently being cited for violation of pedestrian rights. The effects on pedestrian injuries and fatalities is yet to be determined.

Other regulatory approaches to reducing the severity and fatality of pedestrian injuries are through modification of the motor vehicle design. The majority of injuries occurring in a pedestrian-motor vehicle collision are due to the pedestrian being "run under" and contacted only by the wheels and pavement. Approximately 80% of these occur when the pedestrian strikes the front of the vehicle (Ashton, 1982). Ashton estimates that design changes could reduce the number of pedestrians seriously injured to about one-third. This appears overly optimistic to me; however, if this figure is applied to the 50% of pedestrians who are killed at speeds of less than 35 mph (Mueller, Rivara and Bergman, 1988), an estimated 1300 lives might be saved each year.

## Burns

Perhaps the most difficult task I am called on to deal with as a physician is taking care of children with burns. In this injury, the fatality statistics truly underestimate the magnitude of the problem. The 5600 fatal injuries in 1985 are dwarfed by the life-long scars in the 54,000 people hospitalized with burns that year. In addition many of the causes of fatal fires and burn injury differ from those which result in hospital admissions. Burns, I believe epitomize the multitude of interventions which will be necessary to reduce disability and death from this problem. Regulation and legislation have been successful in reducing problems such as children's sleepwear related burns and tap water scalds. Many other burns are also amenable to prevention through regulation/legislation.

One of the most effective of such measures would be the requirement that cigarettes sold in the U.S. be fire safe. Cigarettes that ignite furniture or mattresses kill more people than any other single cause of fire death in the US. The cigarettes are both technically and commercially feasible. This single change would save an estimated 1500 lives and 4600 injuries as a result of the prevention of 230,000 cigarette fires.

Smoke detectors are probably the single cheapest investment a family can make in protecting itself from injury. Liz McLoughlin has shown us that laws requiring smoke detectors in all homes can increase the number of working smoke detectors in homes and decrease the number of fire fatalities (McLoughlin et al., 1985). Building codes should be extended to require all new homes to have smoke detectors hard wired and thus not dependent on yearly battery changes. If all homes had working smoke detectors, 25% of residential fire deaths or 1400 lives per year could be saved.

Perhaps the most significant building code change to decrease fire and burn deaths and injuries would be to require residential sprinklers to be standard in all new construction. The number of deaths prevented through such a requirement is unknown, but is likely to be substantial with time.

The Flammable Fabrics Act was one of the first examples of legislative approaches to injury control. It has made sleepwear related flame burns a thing of the past. However, all too often children with flame burns from the ignition of other clothing are still admitted to burn units with very deep burns, which always require grafting and often result in substantial long term disability. Why not extend the scope of the Flammable Fabrics Act to include all children's clothing? Technically, it is feasible. Although the number of lives saved may not be large, the amount of human suffering this action would prevent is immeasurable.

With the concern over the last two decades on the quality of our air, many communities now regulate outdoor burning. An important source of burn injuries are flame burns from outdoor fires, to which gasoline is often added to accelerate the fire resulting in explosions, clothing ignition and serious burns. Injury control practitioners should combine forces with ecologists to ban all outdoor burning.

## Drowning

Drowning causes over 6000 deaths a year in the U.S., with children and adolescents accounting for one-third of the total. Among children under the age of 5, 60-90% of drowning occur in residential swimming pools. This amounts to 450-650 deaths each year. Essentially all of these deaths can be prevented through adequate fencing which completely surrounds the pool.

Drowning in older children, adolescent and adults obviously have a different etiology. Various studies indicate that one-half of deaths in adolescents and adults involve alcohol. Some of these deaths involve drinking while operating a boat. Control of these deaths and injuries is very difficult. Other boating deaths involve inexperience, lack of training and inadequate equipment. Should we license boat operators and inspect boats as we do now motor vehicles?

## Firearms

Each year in the U.S., 32,000 people die from firearms. Approximately 39% of these firearm deaths are homicides, 56% are suicides and the remaining 5% are unintended shootings. What legislative or regulatory solutions to this problem might one suggest?

The most effective solution is also the least likely to happen: ban all guns. Short of this, one could ban all handguns. Handguns are responsible for 80% of firearm homicides and 65% of firearm suicides. What is likely to be the effect of such a ban on the total number of homicides and suicides in this country? Certainly, there will not be reduction of violent deaths by 22,000, the number of hand gun fatalities each year in the U.S. Perhaps the best data for estimating such as effect comes from our studies in Seattle. Comparison of Seattle to Vancouver B.C., where hand guns are strictly regulated and are thus uncommon, provides at least an estimate of what would be the effect on homicides and suicides if handguns were eliminated in the U.S. Homicides in Seattle were found to be 60% higher than in Vancouver; this difference was accounted for solely by a six-fold increase rate of homicides by handguns in Seattle. Homicides by other means such as knives were identical in the two communities (Sloan et al., 1989). The effect on suicides was not so clear (Sloan, 1990). In the two communities the overall rate of suicides was identical. However, while Seattle had more suicides by firearms, Vancouver had more suicides by other means. Among 15-24 year olds, there was a higher rate of suicides in Seattle, a difference due to a higher number of firearm suicides in Seattle. Thus, eliminating hand guns may result in a decrease of 12,400 homicides annually in the U.S., but have little overall effect on suicides.

The Brady bill currently in Congress would mandate a seven-day waiting period before handgun purchase and require background checks. However, the effect on firearm injury and mortality is likely to be small. Most shootings are not committed by felons or mentally ill people, but by individuals in an act of passion using a handgun which has been previously purchased for home protection. Waiting periods and background checks will have no effect on these events. In addition, the task of trying to identify the mentally ill is very difficult; unless one has been committed, it will be impossible to identify these individuals. Background checks will not work for individuals who acquire their guns illegally, and can be easily circumvented by giving false information. In our study of gun ownership, a large proportion of individuals registering for handgun purchase could not be tracked down because of false or misleading information.

Other regulatory options for reducing the toll of gun violence are to mandate modifications in the bullet or the gun itself. Banning ammunition with magnum charges, mushroom tips and jacketing will not decrease the incidence of firearm violence but might reduce the number of resultant fatalities. The overall effect of this is unknown.

Modification of the weapon itself is also possible and might include changes such as limits on muzzle velocity and automatic trigger locks. Most of these modifications would be aimed at reducing unintended shootings, the number of which is very low compared to firearm suicides and homicides.

## EMS and Trauma Systems

Finally we should not forget about the post injury events, using Haddon's matrix to study injury control. Emergency Medical Systems (EMS) and trauma care systems have been shown to be very effective in decreasing deaths to the 50% of people who do not die immediately at the scene. West and his colleagues have surveyed trauma systems nationwide and found that approximately one-half of the population is not covered with an organized regional trauma care system (West et al., 1985). The proportion of preventable deaths found in unorganized systems varies from 1-71% in various studies (Rivara et al., 1989). Taking a conservative figure of 10% and applying it to the 35,000 people who die in hospitals in states with unorganized trauma care would give an estimated 3500 lives saved with the establishment of an organized system of trauma care nationwide.

## References

Anderson PA, Rivara FP, Maier RV, Drake C. The epidemiology of seatbelt-associated injuries. *J Trauma* 1991; 31.

Ashton SJ. Vehicle design and pedestrian injuries. In A.J. Chapmen, F.M. Wade, H.C. Foot (Eds.); Pedestrian Accidents. Chicester: John Wiley, 1982.

Baum HM, Wells JK, Lund AK. Motor vehicle crash fatalities in the second year of 65 mph speed limits. *J Safety Res* 1990; 21:1-8.

Chorba TL, Reinfurt D, Hulka BS. Efficacy of mandatory seat-belt use legislation. The North Carolina experience from 1983 through 1987. *JAMA* 1988; 260:3593-3597.

Evans L. Restraint effectiveness, occupant protection from cars and fatality reductions. (General Motors Research Publications GMR-6398). Warren, MI: General Motors Research Laboratories, 188.

Haight FA and Olsen RA. Pedestrian safety in the US: some recent trends. *Accid Anal Prevent* 1981; 13:43-55.

IIHS. Superbikes: twice the danger of street cycles. *Status report 1987*; 22:1,4.

Lestina DC, Gloyns PF and Rattenbury SJ. Fatally injured occupants in side impact crashes. *IIHS*, 1991.

McLoughlin E, Marchone M, Hanger SL, German PS, Baker SP. Smoke detector legislation: its effect on owner-occupied homes. *AJPH* 1985; 75:858-862.

Mueller BA, Rivara FP, Bergman AB. Urban-rural location and risk of dying in a pedestrian vehicle collision. *J Trauma* 1988; 28:91.

National Highway Traffic Safety Administration. Fatal Accident Reporting System. 1985. DOT HS 807 071, February 1987 Washington, D.C.: NHTSA.

National Highway Traffic Safety Administration. General Estimates System, 1988. DOT HS 807 607, August, 1990. Washington, D.C.: NHTSA.

National Highway Traffic Safety Administration. Preliminary regulatory impact anlysis: new requirements for passenger cars to meet a dynamic side-impact test. Washington, D.C.: NHTSA, 1988.

National Highway Traffic Safety Administration. Occupant protection facts. June, 1989.

Nichols JL, Ross HL. The effectiveness of legal sanctions in dealing with drinking drivers. In, Surgeon General's Workshop on Drunk Driving. Background papers. US DHHS: Washington, D.C. 1989.

Rice DP, MacKenzie EJ and Associates. Cost of Injury in the United States: A report to Congress. San Francisco, CA: Institute for Health and Aging, University of California and Injury Prevention Center, The Johns Hopkins University, 1989.

Rivara FP, Dicker BG, Bergman AB, Dacey R and Herman C. The public cost of motorcycle trauma. *JAMA* 1988, 260:221-223.

Rivara FP, Maier RV, Mueller BA et al., Evaluation of preventable deaths among pedestrian and bicyclist fatalities. *JAMA* 1989; 261: 566-570.

Robertson LS. Crash involvement of teenaged drivers when driver's education is eliminated from high school. *Am J Public Health* 1980; 70:599.

Robertson LS. Injuries: Causes, control strategies, and public policy. Lexington, MA: Lexington/Heath, 1983.

Sloan JH, Kellermann AL, Reay DT, et al. Handgun regulations, crime assaults and homicide: A tale of two cities. *New Engl Med* 1988; 319:1256-1262.

Sloan JH, Rivara FP, Reay DT, Ferris JAJ, Kellermann AL. Firearm regulation and rates of suicide: A comparison of two metropolitan areas. *New Engl J Med* 1990; 322:369-373.

Voas RB, Rhodenizer AE, Lynn C. Evaluation of Charlottesville checkpoint operations. Technical report. DOT Contract DTNH-22-83-C-05088. Washington, D.C.: NHTSA 1985.

Wagenaar AC, Maybee, Sullivan KP. Mandatory seat belt laws in eight states: a time series evaluation. *J Safety Res* 1988; 19: 51-70.

Wells JK, Williams AF, Fields M. Coverage gaps in seat belt use laws. *Am J Public Health* 1989; 79: 332-333.

West JG, Williams MJ, Trunkey DD, Wolferth CC. Trauma systems. *JAMA* 1988; 259: 3597-3600.

Williams AF, Karpf RS, Zador PL. Variations in minimum licensing age and fatal motor vehicle crashes. *Am J Public Health* 1985; 73:1401-1403.

Zador P, Moshman J and Marcus L. Adoption of right turn on red: Effects on crashes at signalized intersections. *Accid Acal Prevent* 1982; 14: 219-234.

## Discussion

MS. McLOUGHLIN: Do you think that changing the flammability of fabric will affect children's burns? A lot of these cases are gasoline burns.

DR. RIVARA: I think that what happens is a flashback; the kids are standing next to a garbage fire and somebody adds kerosene or gasoline to make it burn better. It flashes up and then the kids' clothes catch on fire. Their clothes aren't soaked with gasoline, the fire flashes up.

DR. BERGMAN: Let me pose a philosophical question. All our presentations presuppose that the government's role is to protect its citizens from injury and to take care of them. Some political pundits term this practice "nannyism." To what extent *should* government extend itself into the protection of people? How far do we go?

MR. MARTIN: That ties into my question about highway safety. One thing that can help save lives is not on the list: maintaining roads and making sure the roads are safe. When the state provides a service, it should be provided in the safest manner possible. That's sound public policy.

DR. WINTEMUTE: On the general question of the government's right or obligation to intervene, we should point out that government picks up the tab for care and has a fiscal obligation, if nothing else, to try and minimize those costs. In addition to Fred Rivara's work on motorcycle injury expenses, there are now two California studies indicating that 85 percent of the firearm injury costs are paid by public dollars.

DR. BERGMAN: What's your response to the question I get asked all the time: "Why don't you pass laws about cholesterol? Cholesterol kills people. Why don't you go after butter and leave us motorcyclists alone?"

MR. TERET: I think we must distinguish between laws designed to prevent people from being harmed by other people and laws intended to prevent people from harming themselves.

The first category of laws finds a lot more public acceptance. Protecting people from harm created by others has been a tradition in the United States and elsewhere for so long that I don't think there is much debate about those laws. The second category - which might include telling people what to eat, to wear helmets while driving their motorcycles or to fasten their seat belts - meets harsher criticism. I think there is good cause for debate about the second category of laws.

MR. MARTIN: From a practical standpoint, we lobbyists know that the farther out from the point of injury, the less dramatic that effect will be on the individual and the harder it will be to make a practical argument in the legislature. There may be some public policy logic to saying "no cholesterol" but legislators are less likely to listen to it.

MS. TRACY: When it came to dollars and cents, for example, on motorcycle helmets, there was a lot of understanding that the multiple millions of dollars that the state was picking up for these injuries was real important. Legislators did pay attention to that.

But I think that the most persuasive argument we used - and I'm sorry it took us so many years to figure it out - was to bring in the spouses and the children of those who had been injured on motorcycles. The committee was persuaded by the fact that it's not just the individual who's being injured, but those who love and depend upon that person. That caught legislators' attention. That was our best argument. Money was important, that's true, but it was during the personal testimonies that I could see the expressions and the mood change.

That may often be true when we play with statistics instead of getting into emotions.

MR. CARPENTER: There are perceptions already at work for us. Whether or not statistics show a waiting period for a gun will reduce fatalities, there is a public perception that some people will go from a bar, buy a Saturday night special and come back to shoot somebody. That perception is there. Statistics may be able to disprove it, but why would you want to disprove that which is advancing your own cause.

There is a perception that if you aren't wearing a motorcycle helmet, you are going to be splattered all over the street. We can build on these ideas without resorting to statistical models which always can be countered by other statistics.

DR. BERGMAN: Whenever I hear politicians asking for scientifically valid statistics I always feel like asking them to point to the important public policy issues that have been decided on the basis of science. Few have been!

# The Effectiveness of Legal Countermeasures Against Alcohol-Impaired Driving

Allan F. Williams, Ph.D

Efforts to reduce motor vehicle injuries need to be guided by the findings of objective, scientific research. Injuries do not receive the priority accorded to other leading health problems, and funds available to reduce motor vehicle injuries are limited. Unfortunately, in the motor vehicle injury area much time and money are spent on so-called countermeasures (primarily involving education or training) that scientific research has shown to be ineffective, or even counterproductive (Lund and Williams, 1985; Lund et al., 1986; Struckman-Johnson et al., 1989). It is obvious but perhaps necessary to emphasize that the key to reducing motor vehicle injuries is to adopt policies known to be effective in doing so.

## Importance of Legal Countermeasures

Focusing on legal countermeasures in regard to alcohol-impaired drivers is appropriate. State laws making driving with high blood alcohol concentrations illegal form the cornerstone of all efforts to reduce alcohol-impaired driving. Focusing on alcohol as the drug of concern also is appropriate. Several drugs other than alcohol have been found to impair performance in laboratory skills tasks (Accident Analysis and Prevention, 1976), but even if they are a causal factor in crashes (which has not been reliably established), they are used by far fewer drivers than alcohol (Williams et al., 1985).

## 1980s Reductions in Alcohol-Impaired Driving

Alcohol-impaired driving and resultant crashes decreased during the 1980s. Between 1980 and 1989 the percentage of fatally injured passenger vehicle drivers with high ($\geq$ 0.10 percent) blood alcohol concentrations (BACs) declined from 53 to 40 percent, a 25 percent drop (Insurance Institute for Highway Safety, 1990). Despite this drop, the problem remains large. Moreover, much of the decline occurred in the early 1980s.

Smaller declines occurred in the late 1980s, and for 16-20-year-olds there was an upturn in the percentage of fatally injured drivers with high BACs in 1988 and 1989.

## Strategies for Addressing the Problem

Alcohol is legal and widely available in the United States and the private automobile is the predominant mode of transportation. Thus, some amount of driving after alcohol consumption is inevitable. The results of a national roadside survey conducted in 1986 indicated that on weekend nights about three out of every 100 drivers on the roads in the U.S. have BACs of $\geq 0.10$ percent. Although this constitutes a large group, there are many more drinking drivers on the roads with lower BACs. Among drinking drivers, 69 percent have low BACs (below 0.05 percent) and 20 percent have moderate BACs (between 0.05 and 0.999 percent); only 12 percent have BACs of 0.10 or greater. Because crash risk increases greatly with ascending BACs, most drinking drivers fatally injured on weekend nights have high BACs (86 percent). Only 5 percent have BACs below 0.05, and 9 percent have BACs between 0.05 and 0.999 (Williams, 1989). These data suggest two strategies for controlling alcohol-impaired driving:

> *Concentrate law enforcement efforts on drivers with high BACs.* Because these drivers constitute the bulk of the problem, it is most important to remove them from the roads and reduce their likelihood of repeat alcohol-impaired driving offenses.
>
> *Deter potential offenders from driving in the first place.* It has been estimated that in the absence of special enforcement efforts, as few as one in 2,000 drivers with high BACs are arrested. Because it is not possible to apprehend more than a relatively small proportion of this very large group, the goal should be to keep drivers with an elevated crash risk (both those with high BACs and those below 0.10 percent) from driving.

### How Laws Can Deter Behavior

The effectiveness of laws against alcohol-impaired driving depends especially on how they affect the perceived likelihood of detection and arrest, and also on the swiftness and the severity of the punishment. In general, sanctions will be effective to the extent that people think it likely that they will be detected and apprehended if they drive while impaired by alcohol, and that, if they are apprehended, punishment will be certain, swiftly applied, and have meaningful consequences. Note that the key to effectiveness is not the law itself, but the way in which it is applied, particularly in terms of the amount of enforcement and the publicity surrounding the enforcement effort. Thus, the same law can be effective in one jurisdiction, but not in another. Or it can be effective at one time, but not at a later time in the same jurisdiction.

### Which Types of Sanctions Work Best?

There is insufficient evidence concerning the effectiveness of fines, and evidence regarding the deterrent effect of jail sentences is inconsistent and inconclusive (Nichols and Ross, 1990). However, license suspension is effective in reducing the subsequent crash involvement of drivers convicted of alcohol offenses; this effect typically continues even after the suspension period has elapsed. Many suspended drivers do drive, but they tend to drive less and, perhaps, more carefully so as to avoid apprehension (Ross and Gonzales, 1988).

### Administrative License Suspension Laws

Historically, in the United States, license actions have been applied after conviction in the courts under criminal procedures. However, in the 1980s about half the states enacted administrative license suspension (or revocation) laws. License suspension applied through administrative procedures rather than through the criminal system is much more efficient. First, the license action is triggered by failure of, or refusal to take, a chemical test, not by criminal conviction; anyone arrested is immediately subject to license suspension or revocation. Usually, the arrested officer confiscates the driver's license and issues a notice of administrative license suspension or revocation, which serves as a temporary license. The driver has a period of time during which a hearing can be requested and its outcome appealed. Hearings are strictly limited in scope. The arresting officer's report is evaluated to determine that there was probable cause to request a chemical test and that the result of that test exceeded the administrative BAC threshold.

### Effectiveness of Administrative License Suspension Laws

These procedures provide swift and sure application of a meaningful penalty. Therefore, administrative license suspension laws, in addition to reducing crash involvement of suspended drivers, have great potential for deterring alcohol-impaired driving in general. Research evidence from single- and multi-state studies indicates that administrative suspension laws do have a general deterrent effect. Two of three major multi-state studies have found that administrative license suspension laws reduce crashes that are likely to be alcohol-related (Zador et al., 1989; Klein, 1989). In one major nationwide study, nighttime fatal crash involvement was estimated to be reduced by about 10 percent as a result of administrative license suspension laws (Zador et al., 1989). Another recent study found no effect of administrative suspension laws, but it is open to criticism because its statistical model did not provide an adequate account of all factors related to changes in motor vehicle fatalities during the study period and thereby produced biased results (Evans et al., 1990).

The weight of the evidence clearly indicates that administrative license suspension laws reduce alcohol related crashes, and a large coalition of 32 highway safety organizations is vigorously promoting their enactment in the 21 states that presently do not have them. Opposition to passage has come from trial lawyers, some labor groups (because a small proportion of suspended drivers lose their jobs), and from those who argue that state costs to administer the laws will be onerous. Research indicates, however, that states achieve cost savings when they implement administration license suspension laws (Lacey et al., 1990).

## Variation in Administrative License Suspension Laws

Now that the majority of states have administrative license suspension laws, a question of interest is whether some of the laws are more effective than others. This is likely because the laws vary in some significant ways (Williams et al., 1990). For example, the swiftness with which license suspension takes effect, barring a hearing, varies between 7 and 45 days. The suspension period for first offenders ranges from 30 days to one year. Suspensions that are in force for longer periods are more effective than short suspensions. There is some evidence that suspensions as short as 30 days may not even be effective during the time they are in force (Nichols and Ross, 1990).

States also differ in administrative procedures and policies that affect the "cost" to a police officer of making an arrest, the likelihood that the arrested party will request a hearing, the results of the hearings, and the likelihood that suspended drivers will have their licenses reinstated once the suspension period has ended. These factors and more are likely to produce state-to-state variations in effectiveness.

An important next step is to determine which administrative laws are most effective in reducing crashes and why.

## Enhancing the Deterrent Effects

The deterrent effects of administrative license suspension laws can be enhanced through the use of techniques that make it more difficult for those with suspended licenses to drive (e.g., by confiscating license plates or vehicles or imposing breath-alcohol interlocks) or easier for police to detect them if they do drive (e.g., special markings on license plates).

Employment of techniques to maximize the effects of license suspension would likely also increase the number of arrests. Many police officers are not enthusiastic about taking licenses because they believe (correctly) that many of these people continue to drive with impunity. Increased and more effective enforcement presumably would reverberate throughout the system and encourage police officers to arrest more alcohol-impaired drivers.

## The Importance of Enforcement

Administrative license suspension laws also can be enhanced (as can any law) through vigorous enforcement that is visible and widely publicized. Creating a public perception that there is a high likelihood of offenders being detected and arrested is the essential component for successful general deterrence. In Nevada, nighttime crashes were reduced when an administrative license suspension law was implemented. Subsequently, a public information campaign emphasizing license revocation and strict enforcement greatly increased awareness of the sanction and expectations that it would be applied. There was a significant further decline in nighttime crashes (Lacey et al., 1990).

Although sobriety checkpoints were not used in the Nevada program, they are effective in promoting general deterrence. Checkpoints are highly visible and they generate considerable publicity about alcohol law enforcement.

## Enforcement of Other Laws

Well-publicized enforcement is important to all laws aimed at alcohol-impaired driving. For example, enforcement will be key in determining the wisdom of reducing the threshold defining impairment (typically 0.10 percent BAC). Several states have reduced the threshold to 0.08 percent.

The American Medical Association has recommended a BAC threshold of 0.05 percent; others have advocated lower thresholds and even zero BACs.

As discussed earlier, we most want to remove from our highways the high BAC group that constitutes a small minority of drinking drivers but the vast majority of those who are fatally injured. If lowering the BAC threshold shifts focus away from the high BAC group, it may limit the effectiveness of the policy. Research evidence on this question is not yet sufficiently available.

Enforcement also is important in regard to minimum alcohol purchase age laws. Now in force in all 50 states, these laws were responsible for a larger 1980s decline in the percentage (10-15 percent) of fatally injured drivers with high BACs among 16- to 20-year-olds than in any other age group. As noted earlier, this was the only age group showing increased fatalities in recent years. This may be because it is becoming easier for underage people to buy alcohol. Findings from an Insurance Institute survey indicate that underage persons can routinely buy a six-pack of beer in retail stores. In Washington, D.C., underage buyers were sold beer in 97 out of 100 stores (Preusser and Williams, 1991).

## Conclusion

State laws making driving with high BACs illegal are an important way to deal with the problem of alcohol-impaired driving. Administrative license suspension procedures based on such laws have been found to be particularly effective in reducing alcohol-related crashes. However, great variations in the 30 existing state laws mean that some are likely to be more effective than others. The key to the success - or lack of success - of a law is not the law itself but how it is applied, particularly in regard to its enforcement and the surrounding publicity.

## References

*Accident Analysis and Prevention.* 1976; 8: 1.

Evans WN, Neville D, Graham JD. General deterrence of drunk driving: Evaluation of recent
American policies. *Risk Analysis,* 1990; in press.

Insurance Institute for Highway Safety. Alcohol Fact Sheet. 1990; Arlington, VA.

Klein T. Changes in alcohol-involved fatal crashes associated with tougher state alcohol legislation. 1989; Brookeville, MD: Sigmastat, Inc.

Lacey JH, Jones RK, Stewart RJ. Cost-Benefit Analysis of Administrative License Suspensions. (DTNH22-88-07310). 1990; Springfield, VA: National Technical Information Service.

Lacey JJ, Stewart JR, Marchetti LM, Jones RK. An assessment of the effects of implementing and publicizing administrative license revocation for DWI in Nevada. 1990; Washington, D.C.: National Highway Traffic Safety Administration, U.S. Dept. of Transportation (DOT HS 807 600).

Lund AK, and Williams F. A review of the literature evaluating the defensive driving course. *Accident Analysis and Prevention* 1985; 17, 449-460.

Lund AK, Williams AF, Zador P. High school driver education: Further evaluation of the Dekalb County study. *Accident Analysis and Prevention* 1986; 18, 349-357.

Nichols JL, and Ross HL. The effectiveness of legal sanctions in dealing with drinking drivers. *Alcohol, Drugs and Driving* 1990; 6, 379-91.

Preusser, David F. and Williams, Allan F. Sales of alcohol to underage purchasers in three New York counties and Washington, DC. 1991; Arlington, VA: Insurance Institute for Highway Safety.

Ross H, and Gonzales P. The effect of license revocation on drunk-driving offenders. *Accident Analysis and Prevention* 1988; 20, 379-91.

Struckman-Johnson DL, Lund AK, Williams AF, Osborne DW. Comparative effects of driver improvement programs on crashes and violations. *Accident Analysis and Prevention* 1989; 21, 349-357.

Williams AF. Alcohol-impaired driving: The problem and some countermeasures. 1989; Arlington, VA: Insurance Institute for Highway Safety.

Williams AF, Peat MA, Crouch DJ, Wells JK, Finkle BS. Drugs in fatally injured young male drivers. *Public Health Reports*, 1985; January-February.

Williams AF, Weinberg K, Fields MM. The effectiveness of administrative license suspension laws. *Alcohol, Drugs and Driving*, 1990; in press.

Zador PL, Lund AK, Fields M, Weinberg K. (1989), Fatal crash involvement and laws against alcohol-impaired driving. *Journal of Public Health Policy* 1989; 4, 467-485.

## Discussion

MS. TRACY: How did you determine the percentage of drivers with 0.05 BACs?

DR. WILLIAMS: They were stopped and tested in a roadside breath-testing survey. There was chemical testing of fatally injured drivers as well. Those are the only drivers who are tested routinely.

DR. BERGMAN: Are sobriety checkpoints a reason that Sweden and similar countries have less drunk driving?

DR. WILLIAMS: Sweden and other countries such as Australia do not have tougher laws than in the U.S., but they enforce them more rigorously, which is why they are effective. In terms of stopping drivers for testing, Australia does things we can't do here because of Constitutional considerations.

MR. CARPENTER: There is a perception that with a 0.10 threshold our average BAC per arrested person is something like 0.15. Are there any studies about the effectiveness of dropping down to a 0.08 threshold?

DR. WILLIAMS: I don't think a trivial change from 0.10 to 0.08 is going to make much difference, but insufficient research has been done to give a definitive answer.

DR. RIVARA: Of the people admitted to Harborview Hospital who have any alcohol in their blood, the average alcohol level is 0.18. Twenty percent are above 0.20, seven percent are above 0.30. Forty-five percent of those admitted with injuries have positive blood alcohols.

MR. GULLICKSON: Is the success of administrative suspension laws based on the deterrence factor or on the repeat offender factor?

DR. WILLIAMS: We know the laws are effective for keeping those who are arrested off the road, and studies have shown that subsequent involvement of that group is lower too. But there also seems to be a general deterrent effect. These studies, which look at nighttime fatal crashes in general, show about a 9 to 10 percent drop and that comes from a general deterrence. So they are effective at both levels, on those they pick up and on those who choose to stay off the roads in the first place

DR. NELSON: The issue may not just be enforcement *per se*, but the perception of enforcement. I don't know that our police can do much more than they are doing now. If people believe there is a high arrest rate or chance of getting caught, they may be more cautious.

DR. WILLIAMS: That's why sobriety checkpoints are so useful. They create a perception of a lot more enforcement than is actually out there.

Our surveys show that while vehicle registration and mileages have been growing, police enforcement resources have been declining. We need to maximize enforcement with these resources.

MS. BODE: In the Sykes case, the United States Supreme Court said sobriety road blocks do not violate the Fourth Amendment. It originated in Michigan, where the State Supreme Court ruled that road blocks were unconstitutional. Why? Because the specific deterrent event - the number of arrests that would be made at the roadblock - were low compared to the rate of arrests an officer would be able to make out on the street.

However, as several states, including Minnesota, argued *in amicus*, that's not why you do sobriety road blocks. Sobriety road blocks are effective for the general deterrent effect through the impact they have on drivers who think they might get caught. The U.S. Supreme Court agreed.

DR. WINTEMUTE: What information is available on the percentage of drivers who continue to drive after an administrative suspension? Among those, if any do, is there evidence that they change their driving behavior?

DR. WILLIAMS: Yes, according to surveys asking people what they've done. Despite the unreliability of self-reporting, these surveys show that most people do continue to drive, but they are a little more cautious about it. That ought to help the cause, but it doesn't keep most of them off the roads entirely.

DR. RIVARA: Their crash rate actually is lower than people who have gone into so-called alcohol rehab programs.

DR. WILLIAMS: Yes, when compared to any kind of treatment, suspension works and the treatment does not.

MR. TERET: Among young drivers who have alcohol involved in their crashes and for whom drinking is illegal, what's known about where they got the alcohol?

DR. WILLIAMS: Virtually nothing.

Many states have passed laws to try to supplement the minimum alcohol purchase age laws. Three or four have a zero or 0.02 BAC for drivers under 21. And they have tried to make the penalties for it more severe.

MR. CARPENTER: Washington does have the mandatory suspension of driving privilege for any alcohol involvement of a criminal nature by any underage person. In other words, a 13-year-old who gets picked up buying beer can't get a driver's license until he's 18.

MR. TERET: While the under-age sales data that you mentioned are fascinating, we shouldn't assume a direct link between young kids involved in alcohol-related crashes and the people who were able to buy from liquor retailers. The people involved in crashes may be getting liquor elsewhere, such as home, or from an older person willing to buy the liquor for them.

DR. WILLIAMS: I agree you cannot make that link based on the data we have. However, we sent people into national chain supermarkets to buy just a six-pack of beer. To me that's suspicious already. Some of these stores have something that pops up on the register that says, ID required; supposedly there's no sale without a check. But the clerks simply override that. They were routinely selling beer to the kids we brought in. I was a little surprised.

# Driving While Intoxicated (DWI) Legislation: The Minnesota Experience

Nancy J. Bode, J.D.

Minnesota's attack on drunken drivers has become a national model. For several years a federal grant has allowed us to assist other states as well as provinces of Canada in passing effective DWI legislation.

## General Background

New York adopted the first implied consent statute in 1953 and, with federal encouragement, Illinois became the 50th state to do so in 1972. Minnesota's statute was enacted in 1961. Like most, it provided that the operation of motor vehicles "implied consent" to submit to tests to determine blood alcohol content when there was probable cause to believe the person had committed a DWI violation. It gave the driver the option (but not the right) to refuse to submit to testing, and provided for a revocation of driving privileges for those who refused to be tested.

## Administrative *per se* (1976)

In the early 1970s, Hennepin County (Minneapolis and suburbs) had a federally-funded Alcohol Safety Action Project designed to increase the number of DWI arrests. It was outstandingly successful. However, failure to provide additional resources to deal with the increased caseload led to a terrible clogging of the court system. This increased the pressure for plea bargaining to the point where about half of all DWI charges in the county were routinely reduced, usually to "Careless Driving," even where the driver had a long history of prior DWI arrests. In 1976, this state of affairs came to the attention of State Senator Alex Olson, a former congressman and later lieutenant governor. His outrage at the extent to which plea bargaining was nullifying the legislative intent to withdraw driving privileges from DWI violators led him to propose what became the nation's first "administrative *per se*" law.

Sen. Olson's proposal was so remarkably simple and obvious that one now wonders why it had not been made much earlier. It was simply to take the procedure already being used to administratively revoke the driving privileges of those who refused testing, and adapt it to those whose tests disclosed an alcohol concentration of 0.10 or more. The suggestion won quick approval, and was enacted in 1976.

## Immediate Seizure of License and Service of Notice of Revocation (1978)

Perhaps the most significant change made in Minnesota's DWI control system occurred in 1978 when every police officer became an agent of the Commissioner of Public Safety for the limited purpose of serving a notice of license revocation.

The idea originated with a legislator concerned about the extent to which persons charged with "Driving After Revocation" were avoiding conviction by merely denying any knowledge of the revocation. Personally serving the notice and order of revocation at the time of the incident assured that there would be a record on file.

An unexpected benefit was the encouragement this procedure gave officers to make arrests. Many officers were reluctant to make DWI arrests because they perceived that the time and paperwork involved outweighed the bother, especially if the charges would be reduced to "Careless Driving" anyway. With the new procedure, officers felt there would be an immediate impact on the driver. More officers made arrests. As the number of arrests and resulting driver license revocations mounted, the death rate dropped.

Unfortunately, the legislation also greatly increased the caseload for the attorney general's office, which handled revocation reviews in district court. (While some states have administrative rather than judicial review of revocations, Minnesota offers both.) By 1982, requests for judicial hearings were being filed at the rate of 1,000 per month, while only 900 cases were being closed.

Looking to other states for solutions to the building caseload backlog, we borrowed from a Massachusetts law which imposed the license revocation sanction before any opportunity for a hearing.

## Prehearing Revocation (1982)

Apart from the need for a change caused by physical overload, several other events came together to facilitate the change. Both the governor and the attorney general, who were expected to face each other in the 1982 gubernatorial race, wanted to do something effective about the DWI problem. Mothers Against Drunk Driving (MADD) began to organize in Minnesota as part of the nationwide tide of outrage and frustration over the continuing death tolls. A horrible example also galvanized public opinion when Danny Roman Kramarczyk, while awaiting trial on DWI (and two implied consent cases), struck and killed two women aged 89 and 85, who were helping each other home from church just before Christmas in 1981. He was driving at dusk, without headlights, with an alcohol concentration in excess of 0.10.

By agreement between the staffs of the governor and the attorney general, the governor proposed changes only in the DWI statute itself, while the attorney general proposed changes only in the implied consent statute. Both proposals were drafted mainly by the attorney general's staff, and there were no conflicts between them. News media, hoping to find controversy, were surprised to find that the two rivals had nothing but praise for each other's bills. Members of MADD strongly supported both bills, and personally contacted each legislator at least twice. The two bills were combined into a single bill in the Senate, and, with a few additions, emerged from the Legislature with a unanimous vote.

The chief DWI amendment made the second offense within five years, or the third within ten years, a gross misdemeanor, punishable with up to a year in jail and $4,000 in fines. The big change in the implied consent statute was the substitution of prehearing revocation for prerevocation hearings.

## Annual DWI Amendments

Since 1982, the Minnesota Legislature has continued to pay annual attention to the DWI problem. In 1983 the "Criminal Negligence" statute was broadened into "Criminal Vehicular Operation." Under that statute, it is a felony to kill or seriously injure someone through grossly negligent operation of a motor vehicle while "under the influence" or with a prohibited alcohol concentration. Hit and run violations involving serious injury or death also have been made felonies.

The catalyst for the 1983 changes was an incident in which a swimming coach struck and killed two young girls walking along a road (one of them a member of his swimming team) and left the scene of the accident. When tracked down several days later, he denied any alcohol involvement. Investigation of his background disclosed that he had been involved in a half a dozen or more DWI charges in Kentucky, Wisconsin and Minnesota, most of which were unknown to Minnesota authorities. A later case involving a drunk driver who killed a pregnant woman and her unborn child led to the extension of the criminal vehicular operation law to unborn children.

A federally funded DWI Task Force focused upon the problem of repeat DWI offenders who drive even though their licenses have long been revoked. The task force's proposal for the impoundment of license plates for repeat offenders was passed in 1988. While the bill was being considered, one of the newspapers reported on a repeat offender who killed someone while driving drunk without a license. (News stories about repeat drunk drivers who consistently drive without licenses are fairly common. A local television station even filmed a DWI offender driving after the license was revoked. MADD has been responsible for a vast majority of these stories.) Legislation to remove the plates of a vehicle which is driven by a person who is undeterred by a driving-after-revocation charge or an additional DWI charge was adopted almost without dissent.

In 1989, the media widely reported a Minnesota Court of Appeals decision overturning a drunk driver's conviction for criminal vehicular operation where the jury had been erroneously instructed that it could consider the defendant's drunken condition alone to be negligence. The 1990 legislative session overturned the decision by permitting an alcohol concentration of 0.10 or above to be the basis for a criminal vehicular homicide charge without an additional element of negligence.

Reacting to federally imposed mandates, the 1989 Legislature passed a 0.04 level of alcohol concentration for operators of commercial motor vehicles. The legislation sailed through the the same session in which issues surrounding drugs and alcohol in the workplace were vigorously debated. This was one "drugs in the workplace" bill all legislators could agree upon. Recent publicity over Northwest Airlines pilots who flew a commercial flight after a heavy night of drinking caused the Legislature to set a 0.04 alcohol level for pilots in their workplace.

**A Lesson in Politics**

Some DWI legislative initiatives are unsuccessful. For example, in 1989 a proposal to reduce the legal alcohol limit from 0.10 to 0.08 was defeated in the Senate Judiciary Committee by a 7-8 vote. The defeat was probably a result of several miscalculations. The bill's author, a member of the House of Representatives, originally introduced the bill to reduce the limit to 0.05. The 0.05 legislation was announced with much fanfare, but without prior notice to some of the key players in previously successful DWI bills. The author was seen by some as a person who wanted headlines more than effective DWI legislation. There was no tragedy to which he was responding, no specific problem that he was going to solve.

The bill was amended to 0.08 in order to obtain a Senate author, but the Department of Public Safety, the Attorney General's Office and the DWI Task Force were unprepared to provide sound support for the selection of the 0.08 level (let alone the 0.05 level). Legislators reacted angrily to the suggestion that an 0.08 level was needed to insure prosecution at the 0.10 level. (A prosecutor who admitted that his office routinely plea bargained to "careless driving" all first-time offenders testing 0.12 or more was sharply criticized by the Senate Judiciary Committee.) Finally, the hospitality industry made its first public appearance and testified against the 0.08 legislation. (While a legislator cannot be in favor of drunk driving, he or she may support a person's "right" to have a drink or two.) As a result, legislation that could have succeeded was narrowly defeated.

## The Result

Whether reacting to media reports of tragedies, appellate decisions or attorney general or DWI Task Force initiatives, legislators are reflecting increased public awareness about the dangers of drinking and driving. "There but for the grace of God go I" is a less common attitude in the Legislature than once was the case. Ad campaigns urge those who drink to avoid driving. Designated drivers are becoming common. SADD is joining MADD as a voice against the dangers of drinking and driving. "Sober cabs" are promoted and used on holidays traditionally associated with drinking. Sobriety roadblocks add to the perception that drinking drivers will be caught.

Boating while intoxicated laws follow our drunk driving laws; while there's no license to be revoked, the privilege of boating on one of Minnesota's lovely 10,000 lakes can be withdrawn. Media reports about newly enacted DWI legislation enforce the belief that loopholes no longer exist, and harsh penalties await those who are caught. A number of well-publicized arrests of public figures, including politicians, television personalities, judges, police officers and clergy, have enhanced the perception that drunk driving laws will be applied to any driver, regardless of status.

By building strong coalitions and presenting sound legislative proposals aimed at solving specific problems, Minnesota has acquired a solid statutory arsenal to combat the drinking driver. As a result, Minnesota has seen a decrease in the rate of alcohol-related traffic fatalities. Minnesota has had one of the lowest traffic death rates in the country for four of the past five years. We can only hope that we are able to find even more effective methods to prevent the anguish suffered by victims of drunk drivers.

## Discussion

MS. McLOUGHLIN: Is Minnesota's drinking and driving record significantly better than that of other states?

MS. BODE: Yes. Three of the last five years we were the safest state in the nation and we then tied for third. We have significantly reduced our alcohol-related traffic fatalities and traffic fatalities as a whole.

DR. BERGMAN: I notice you do not mention physicians or insurance companies as allies.

MS. BODE: We have never had physicians or insurance companies come in on these drunk driving bills. I don't know why, except that the legislature is accustomed to dealing with certain players.

MR. FERGUSON: You talked about the hospitality industry's influence. Do you have a meetings law and public disclosure law for officials in that state?

MS. BODE: We have some public disclosure laws, which is how we found out about meetings.

DR. BERGMAN: Do our four lobbyists have reactions or comments?

MS. TRACY: While I've worked on a lot of other prevention bills, this is my first year on alcohol. In hearings where there are any number of statistics, I watch eyes glaze. In this, as in other prevention bills, I feel very much it has to be the personal approach that Nancy was speaking of.

MR. MARTIN: Washington State's trial lawyers have not taken any position on 0.08 or 0.04 for minors or 0.00 for minors. Nor have we taken a position on sobriety checkpoints, although a number of my members have strong feelings on the subject.

My members have focused on the administrative revocation question, so I'll give you a quick overview of their concerns. When a driver is apprehended, there's really one concern: Am I going to lose my license? Many times that is the only thing an attorney, or anyone else, can use to make an individual deal with a drinking problem. Alcoholism is a disease and the problem will continue unless the client gets treatment. That's a deferred prosecution approach.

Our members are starting to look for ways to improve deferred prosecution. The courts are full, prosecutors and judges are overloaded, the system of testing, at least in this state, is not considered completely reliable. (I would be interested in seeing how administrative revocation overlaps with questions of blood alcohol vs. breath alcohol tests and debates about flexibility on the reading.) If you have a 0.00 standard and an unreliable testing system, innocent people are going to be prosecuted. Other people will, in effect, use the system by being picked up repeatedly and getting out for careless or negligent driving on a non-alcohol-related offense.

One thing we're looking at for deferred prosecution is creating a lesser alcohol-related offense, so that a first-time alcohol-related offense is on the record. Our association is in no way saying that society shouldn't deal with the repeat offender, but there should be one opportunity to deal with the treatment issue and try to get rid of the drinking problem forever.

MS. BODE: We allow limited licenses freely to go to chemical dependency treatment and on a first offense we allow a limited license for work purposes.

DR. BERGMAN: Have you negotiated with trial lawyers?

MS. BODE: No, the trial lawyers in our state mostly represent the plaintiff bar. They came in very hard, and we supported them, with social host liability for underage drinkers. But the defense bar has been amazingly silent on the majority of DWI legislation.

MR. TERET: In Maryland a number of years ago we had a politics versus science situation in which a rehabilitation/treatment program was highly touted for political reasons. The governor's task force on drunk driving favored this program, which involved sending people to Alcoholics Anonymous plus additional measures. It was administered by a judge who was a reforming alcoholic, as were his political allies on the governor's task force.

A program evaluation requested by the state revealed that the recidivism rate of those who went through the program was the same as the recidivism rate of those who received no treatment at all. The state withheld that information for some time, then released it between Christmas and New Years when it was felt that people wouldn't be reading the newspapers. Despite the findings, the program was implemented on a grander, statewide scale at a cost of $3 million a year.

It taught those involved in the science part of the program evaluation the gross limitation of epidemiologic data as an influence on the political process. The decision had already been made to implement the program on a statewide basis. The study was a minor temporary embarrassment because the decision was going to be implemented anyway.

DR. BERGMAN: Along those lines I would like to describe our experience last year in the Washington State Legislature with administrative per se.

The bill had the best coalition imaginable. It was pushed aggressively by the governor's office. It was pushed by law enforcement. It was pushed by ourselves, the medical association. MADD, the victims' group, gave very very impressive testimony. It seemed like it couldn't fail, but the bill was absolutely shattered.

We were beaten by the alcohol treatment people. They waged a magnificent campaign. First, they did it in the grass roots by having the centers throughout the state call their representatives and say this would keep people from getting alcohol treatment, that the threat of license revocation is the only thing that will get people into treatment. They were so tremendously successful that they didn't even need to negotiate with us.

This is a huge industry that contributes heavily to political campaigns.

MR. CARPENTER: I believe Washington is the only state where the same agency serves both to evaluate the alcoholism problem for the court as well as provide the recommended treatment. Legislation may be sought to prohibit that obvious conflict of interest.

MS. BODE: Treatment providers are a very potent group and we have been careful not to cross them.

DR. BERGMAN: Has MADD lost political power?

MR. McGUIRE: They have lost a lot of political power in the minds of people who follow them closely. However, we did some polling about a year and a half ago and some follow-ups about eight months ago on how MADD is perceived in California. They have a recognition rate of around 94 or 95 percent, way above the governor of California or anyone else. Way above Paul Newman. Among the general public, MADD's name is perceived as being very powerful and good.

I would say that about 0.001 percent of the population thinks they have become a scam. I'm in that camp, but the rest of population thinks they are apple pie.

DR. BERGMAN: Why do you think MADD is a scam?

MR. McGUIRE: Because they've refused to support tax increases on alcohol at the federal and/or state levels; they have refused to go after equal time for counter-advertising, and they have in the past (although not now) received a significant amount of money from the liquor industry. I think anytime you take on the alcohol issue, whether it's the *per se* law or whatever, the one factor that is never discussed is the liquor industry's role in promoting drunk drivers or homicides or alcoholism by improving and enlarging their base of customers, especially young people. If you're going to look at the politics of all of this, I think you have to look at the industry that supplies the product we are talking about and MADD has not been on the side of taking on the industry.

Now there may be changes because they have received a lot of criticism from that 0.001 percent of the population, so I'm willing to say that the jury's still out.

MS. BODE: MADD rarely testifies on our bills. Instead the major players are the attorney general's office, the DWI Task Force, and the Department of Public Safety. Our office comes in and introduces some of the legal reasons for a bill and how it will work, the department comes in with the numbers, and the DWI Task Force tries to field other questions. We rarely have MADD testify, partly, I think, because its effectiveness has been reduced. However, MADD is very effective at hooking up the media with the victims.

If you want to pass DWI legislation you have to figure out how to use organizations like MADD effectively. Before every legislative session we sit down with the key players and ask "what are you introducing, what are you pushing this session." Then we try to figure out which thrusts we should have in the session. We have had a good working relationship with MADD in Minnesota and they have backed off some of their proposals in order to support other proposals by the task force.

For example, as part of the impoundment bill MADD wanted cars to be confiscated. We got them to support the more limited proposal of impounding plates, and they were very effective representing victims. You need to pull at the heart strings as well as present statistics.

# The Art (and Necessity) of Coalition Building: The California Alcohol Tax Initiative

Andrew McGuire

The essence of running an initiative campaign in California is building a coalition, raising money, and using TV and radio. Of the three, clearly number two is most important. I'd like to review the two-year period of time that I worked on the alcohol tax initiative to give you a sense of what it's like to get something pushed forward on a state level in California.

## Historical Background

In 1981, while I was on the board of MADD, we advocated a nickel-a-drink tax on alcohol at the bar and restaurant, a tipplers tax if you will. The actual language of that bill imposed the tax for a one-year period only, to be used for restitution to victims of drunk driving and for education programs. We all thought at the time that was incredibly reasonable, but the liquor lobby killed it. Although it didn't even get out of the first committee in the Assembly, it introduced me to the notion of taxing alcohol and using that tax money for something worthwhile.

A few years later, in 1986, Assemblyman Lloyd Connolly introduced a bill to increase the taxes on tobacco. That didn't get out of the first committee either, so the next year Assemblyman Connolly began organizing a coalition to put forward an initiative to tax cigarettes. That was the only way to get around the tobacco lobby. In November of '88 the cigarette tax, 25 cents a pack, was passed as Proposition 99. It was on the same ballot as the well-known insurance initiative, No. 103.

I advised and peripherally worked on the cigarette tax initiative because originally it was drafted to allocate some of the tax money for fire prevention. Ultimately that didn't happen, but it got me connected to supporters of the cigarette tax initiative.

A month before the November 1988 vote, I met with Connolly's key circle working on that initiative and said "if 99 passes (and it wasn't clear it was going to) would any of you ever consider doing an alcohol tax?"

They said, "Yes, we would be interested. If it passes let's get together." In December of 1988 I attended the first organizing meeting for the alcohol tax initiative, which was defeated this past November. I ultimately became the chair of the campaign in California.

## Why an Initiative?

There are basically three phrases to any initiative campaign. First is the drafting phase, where you bring together the thinkers and coalition members who deal with the issue and try to create a consensus. Like the tobacco tax proponents, we chose the initiative process because there is absolutely no way to raise alcohol taxes through the California legislature. It had been tried 14 times in the last 10 years, and always failed.

To give you an index of the legislature's attitude, the wine tax in California in 1937 was two cents a gallon. Late that year, it was reduced to one cent a gallon where it's remained for 54 years. At a penny a gallon, the wine tax in California is obviously the lowest in the country. It is basically a non-tax. The highest wine tax per gallon is in Florida at about $2.20 a gallon. The disparity is obvious. The beer tax in California is four cents a gallon. The only two states where it's lower are (Anhueser) Missouri (and Miller) Wisconsin, where it's three cents a gallon. In most states the beer tax is between $.80 and $1.20 a gallon. For distilled spirits we're at $2 a gallon, which is very low. The highest, I think, is Utah at $12.50 a gallon. That all goes to show that California has the lowest alcohol taxes in the nation. And there are 18 lobbyists in Sacramento working to keep it that way.

Going outside of the legislative process is in many respects bad public policy. I know those arguments inside and out: "It doesn't allow enough public input." "It's just a handful of people in the back room drafting the initiative." Sometimes I have a hard time seeing the difference between that and the legislative process, but that's the conventional wisdom. Between December 1988 and October 1989, we built coalitions and drafted the initiative.

## Coalition Building (and Unbuilding)

The coalition building started with bringing in people who work in the alcohol field, including a few alcohol policy researchers, county coordinators for alcohol and drug programs, people from recovery home associations (which are very strong in California and other states) and so on. Shortly thereafter we brought in emergency physicians, who were very interested, and county supervisors, who also were very interested because they have to deal with the alcohol problem as it's exhibited through homelessness, mental health problems or public alcohol issues.

The coalition building went through a frightful process because of the astounding sum of money the tax would raise. Basically this is what happened. When we started discussing where to set the tax, we chose a nickel a drink because in May of '88 then Surgeon General Everett Koop recommended that states increase their alcohol taxes up to a nickel a drink. That became our scientific justification for a 5-cent tax. In California a nickel a drink raises about $750 to $800 million a year. When you are trying to decide how that much money will be spent, into what programs it will go, the coalition builds itself. The real problem was figuring out how to include the right people and then how to delicately and gracefully and graciously exclude others.

I'd never before been in the position tearing down coalitions rather than building them. I'll give you one anecdote. Initially 10 percent of the money was to be earmarked for environmental issues. The loose connection was that there are a lot beer bottles littering the environment, a lot of beer caps and plastic bottles being thrown away, there is a lot of destruction by young drunk males in public parks. That's all true, but earmarking $75 to $100 million a year for the environment from an alcohol tax didn't grab people in our two major statewide polls.

Six months into the campaign, I had to literally kick one of the leaders of the environmental movement in California out of the room. I recommend using this ploy, by the way. I simply said, "Jerry, I don't want to say whether or not you're going to have to leave. I just want you to put yourself in my shoes and tell me what you would tell yourself." Of course, he said, "Well, I'd tell me to leave." And I said, "Well, that's I guess what I'll have to say."

## Drafting

From the beginning, the policy issue was central to me. There is substantial evidence that increasing the price or the excise tax on alcohol reduces consumption, particularly among younger people who have less disposable income. That got me involved; it was worth a significant amount of my time and energy to reduce alcohol consumption which, in turn, would reduce injuries.

Most of the other people involved in the campaign were interested in getting money for their programs. For example, doctors wanted money to fund the uninsured cases that were coming up; treatment centers wanted to stay open or to expand their programs. I probably surfaced as the chair of the campaign because I was the only person representing a group that wouldn't get any money from the initiative. That was an extremely valuable tool to have at my disposal. It gave me a kind of moral suasion over some of these people.

The drafting phase went through 27 drafts. I would say the first 20 drafts were simply going from funding for emergency care, alcohol treatment prevention to other issues that came along. The initiative language grew and grew and grew for 20 drafts until it ended up with five areas of funding: mental health, alcoholism treatment and prevention, trauma care, law enforcement and finally what we called innocent victims (child abuse, battered women's shelters and so on). In the end, everything to be funded had a link with alcohol as a contributing factor to the problem or issue or program.

When we finally finished the drafting process, we ended up with about 13 or 14 people representing 14 different constituencies. That became the campaign committee that I chaired.

### Signature Gathering

After drafting and submitting the document to the attorney general for approval, we moved into the second phase, which was signature gathering. In California and about half the states, it takes a certain number of signatures to put an initiative on the ballot. We had to get eight percent of the number of people who voted for the governor in the last election, about 650,000 valid signatures. So a million signatures was the goal. The time frame was 150 days. Getting a million signatures in 150 days in California is no small feat, I guarantee you. We hired the firm that gathered the signatures for the cigarette tax initiative, and in the end our coalition went out and got about 440,000 of those signatures on a volunteer basis. For the remaining signatures, ultimately over 800,000, we actually paid 50 cents apiece.

### Fund-Raising

By quickly calculating in your head, you can see that a major issue of fund-raising loomed large. Sitting in my office at San Francisco General Hospital where I was, and still am, running a nonprofit organization that's not supposed to lobby, I was in charge of raising approximately $5,000 a day for 150 days for a political activity. There's room for a whole other discussion on how the IRS regulations on non-profits stop major efforts in injury control.

The money-raising issue ties directly into coalition networking, because if you don't have a wide circle of friends and organizations to call, you will never raise the money. Very little attention is ever placed on the skill or ability to raise money among injury control activists because so few of us have ever had to raise big bucks for political issues.

## The Campaign

The campaign itself began after we finished signature gathering. The May 15th, 1990, deadline signaled the beginning of what the general public would perceive as the campaign. In early August, we received a ballot number and became Proposition 134.

Let me briefly review some of the key features of that period of time. Number one, the liquor industry began its negative campaign against us before we even had a number. It started a week after we qualified, making it the earliest negative campaign against an initiative in California history. We know through financial disclosure reports, that the liquor industry raised between $40 and $45 million. We raised $1.3 million. Similarly, the cigarette tax initiative spent $1.6. And the tobacco industry spent $20 million to fight it. By comparison, about $20 million was spent by both sides combined in the governor's race. Ours was the most expensive campaign in the entire country in 1990.

Needless to say the liquor industry had no problem raising money. About $3 million came from California, the rest from out of state. The Beer Institute, from Washington, D.C., dumped in an early $10 million. There was money from Guinness, from Bacardi in Puerto Rico, from Philip Morris, from everywhere. The money flowed in; a whole host of political consultants went to work.

With the campaign by the liquor industry starting so early, we had incredibly false optimistic visions of public awareness working in our favor. Everyone knew there was something about an alcohol tax on the ballot. The advertising started at least five weeks before the second negative campaign started against the environmental initiative known as Big Green.

The liquor industry's early start created poll results showing us with an early lead in the 70/30 range. A month into the campaign we had a little bit of decay in our support. A month before the Nov. 6th vote we were at 65/35 in the polls. Of course as you get closer to any election, you get more and more polls: the L.A. Times poll, a KCRA poll, a Sacramento TV poll. About nine or ten polls in the last month all showed us with between 55 and 65 percent yes votes. One of the lowest percentage of yeses (55/31) was in Orange County, an area that never supports anything like this. So a month out, it looked like a repeat of Proposition 99.

Two weeks out, our final field poll showed us down to about 60/35 and we were pleasantly surprised. Then one outlier poll, which we internally commissioned, showed us below 50 percent. Our consultant, whom we trusted said, "I don't understand it, it just happened in the last week." In his polling for numerous other clients, he noticed that something equally bizarre had happened 10 days out.

**The Last 10 Days**

This is what happened: California voters began to focus on 28 initiatives on the ballot, a new insurance commission position, statewide offices, the governor's race, all the congressional races and so on. Anger set in during those last 10 days. With a couple of exceptions, basically everything on the ballot lost, including us. As we look back, we see that we were dragged down by the voters' anger and frustration.

Our state ballot was, in total, 220 pages. We're talking 220 pages of fine print, pages that the elderly and others without good eyesight cannot read. The 220 pages, the 28 initiatives (Big Green alone took up 42 pages of the ballot book) were more than the people would accept.

More importantly, we had also negative initiatives. We had a negative initiative against Big Green, we had a negative initiative against an initiative to save redwood trees. The alcohol tax initiative had two: one negative initiative and another that required a two-thirds vote for us to win. Altogether the voters could not understand what was what and just said no. Again, I want to emphasize that this occurred in the last 10 days, when the voters focused on the upcoming election.

## What We Gained

The campaign process, as seen from mid-May to Nov. 6th, showed the strengths and weaknesses of the coalitions that were built early on. It also showed that "the alcohol field" is in its infancy when it comes to knowing how to be political, how to work in the trenches, how to get people to appear at a press conference, how to place a proper well-informed spokesperson on a local radio talk show and so on. There was not only coalition building, but simultaneous, in-service training going on.

Now many of the coalition's alcohol constituents (and including many many people involved in the drunk driving issue, excluding MADD) are looking back on the whole campaign as the milestone that galvanized them and will now push them to a new level. In the future, they expect to be more effective in taking on their own separate issues and in learning how to work with one another. It was a trial by fire, if you will, for a lot of these groups. We set up coalitions in the 15 most populous counties in California. Members were, in a sense, yanked out of their everyday jobs as a county administrators dealing with alcohol programs, as leaders of child abuse councils or whatever, and forced to learn the importance of mastering the political process, raising money and reaching the media.

The artificial experience of the initiative campaign probably catapulted forward the political expertise among alcohol-related groups that will deal with alcohol problems in the future. The main agenda item for all these groups will be dealing with advertising of alcohol products.

One footnote: Gov. Pete Wilson's 1991 budget included a proposal to raise the alcohol tax. You lose the battle and win the war. That formerly was unthinkable for any California governor, let alone a Republican. I must point out that we're facing a $6 to $9 billion deficit in California and they have to get money somewhere. Even so, proposing a potential alcohol tax increase was politically non-viable two years ago and now it's the cornerstone of one of Wilson's tax increases.

## Discussion

MR. FERGUSON: One of the few initiatives that passed in 1990 in California was a limitation on terms for legislators. Why do you think the voters distinguished that issue from all the others that went down to defeat?

MR. McGUIRE: I was familiar with some of the polls on that. The voters angry with the initiative process, and angry over trying to figure out whether Wilson and Feinstein were really different were looking for an outlet. When 140 (to limit the terms of legislators to six years) popped up, it was just what the angry voters were looking for.

MR. FERGUSON: Didn't it have tremendous money against it?

MR. McGUIRE: Yes, and a third of the money (about $5 or $6 million) was from the liquor industry. Why? Because the leader of the campaign against the term limitation was Speaker of the Assembly Willie Brown.

There was a significant *quid pro quo* with the liquor industry because he was also against our initiative.

MS. TRACY: Why would a nickel a drink keep anybody from drinking?

MR. McGUIRE: We never, ever said that raising the tax would decrease consumption because no one believes it. But the research shows it. Whenever a state or a country raises the tax, consumption goes down.

For example, when California raised cigarette taxes 25 cents a pack, there was a 14 percent decrease in consumption after the first year. The rest of the nation had about 3 1/2 to 4 percent decrease in consumption. California was the only state with an extra 10 percent decrease in consumption, and the only variable was the 25 percent per pack increase. It tends to show up in people who have less disposable income, usually young people.

DR. BERGMAN: Would the tax have provided direct money to launch things like counter commercials?

MR. McGUIRE: Yes. The part of the initiative that I actually wrote earmarked 2 percent of the money (about $15 million a year) to do counter advertising on TV and radio against alcohol promotion. I know from off-the-record talks with people from the other side that was what they feared the most, not the increase in the price.

DR. RIVARA: Who were your contributors?

MR. McGUIRE: Our largest private personal donor simply wrote us a check for $60,000. The next largest private donor gave $5,000. Hundreds of people gave us $50 to $500. Our direct mail appeals didn't raise much. The emergency physicians went out and raised $240,000 among their members. The battered women's shelters raised $30,000, and child abuse councils raised $45,000 from their members.

Treatment programs contributed a couple of thousand. Keep in mind that the alcohol treatment money to be raised by the tax was earmarked for county programs, not private. The private folks were horribly upset that we weren't putting money into the private sector, so they weren't about to raise money.

MR. TERET: Do you think that the response of the alcohol industry would have been less vigorous, if you hadn't included the provision about advertising?

MR. McGUIRE: No. To put this in a national context, simultaneously a federal tax increase on alcohol was going through Congress. We don't think it had a huge impact on us, but it did pass 11 days before our vote. We did some initial polling after the election, and most people hardly noticed that it had happened. It hadn't actually taken effect so most people weren't aware of it.

# A Campaign to Curtail Irresponsible Alcohol Advertising in Washington State

Abraham B. Bergman, MD

Our campaign here in Washington State, led by the Washington State Medical Association (WSMA) has been directed towards curtailing the irresponsible advertising of alcohol, mostly beer, directed to the group that is at highest risk for alcohol-related death, 16-25 year olds. We have eschewed moral arguments, attempting instead to make alcohol advertising a health issue. We are not opposed to responsible drinking nor responsible advertising of alcohol. We do object to the blatant attempts of the beer companies to equate consumption of their product with "the good life" through use of sex, rock music, athletic prowess and other risk taking behaviors.

Attempts continue to be made in Washington, D.C. to curtail alcohol advertising. There are current efforts in Congress to mandate warning labels on alcoholic beverages. That effort does not excite me one bit. There is no evidence that warning labels have any effect on consumption. The battle consumes a great deal of energy that could expended elsewhere, and the result is marginal even in the unlikely event that the proponents prevail.

Also, because of the firepower of the opponents, I do not think it is possible to get meaningful alcohol reform legislation through the United States Congress. We are not only talking about the alcohol industry. We face every advertising agency in America, the hospitality industry, and most powerful of all, newspapers, radio, and television. Those are powerful enemies.

Our feeling therefore is that since little can be done in Washington, D.C., the battle against alcohol advertising practices must take place on a state level. To this end we are aided by a curious historical anomaly, the 21st Amendment to the United States Constitution. When Prohibition was repealed, several states, especially in the South, wanted it retained. In order to get the votes of those states, the language of the Amendment reserved the rights of states to maintain control over the commerce of alcohol. Therefore, the interstate commerce clause that restricts the rights of states to interfere in the commerce of products that cross state lines, does not apply with the same degree of force to alcohol.

We feel, therefore, that given the technology of communication and advertising where the states are so interdependent, the actions of one state in forcing modification of advertising practices would have a national impact. We are not the only ones who think so. At a public hearing before the Washington State Liquor Control Board last November 28th, the room was full of industry lawyers and public relations flacks who had flown out from New York and Washington, D.C. to influence us yahoos.

## Constitutional Issues

There are 18 States directly involved in the sale and distribution of liquor. In Washington this activity is regulated by a 3 person board appointed by the governor. Regulation of advertising is one of the Board's statutory responsibilities. Citing alcohol-related trauma as the leading cause of death between ages 16-24, the WSMA petitioned the Liquor Board to curb "irresponsible" advertising directed toward that vulnerable group. With the knowledge that if the Board granted our petition, an immediate legal challenge would be launched that might even go to the U.S. Supreme Court, the document was drafted with great care by the WSMA attorney, Greg Miller. Time does not permit a detailed discussion, but in addition to the "commerce clause," these were some of the other constitutional issues and how they were addressed.

### Vagueness of proposed regulation

We proposed the voluntary advertising code of the Wine Institute as the model to be followed (no sexual themes, rock music, recognized celebrities, party-time, etc.). If advertisers had difficulty in understanding that code, they could then employ "tombstone advertising," which is perfectly clear. "Tombstone advertising" means black and white, no music, and no human figures.

### Censorship

The ads themselves would not be censored. Rather the product of any company that violates the advertising code could not be sold in the state.

In 1980 the U.S. Supreme Court differentiated between individual free speech, and commercial free speech, which can under certain conditions be restricted. (*Central Hudson v. New York 1980*). The restriction can be applied if it is based on a "substantial governmental interest," which in our case, is control of alcohol-related trauma. Obviously the free speech issue is more complex and will undoubtedly continue to attract attorneys arguing vehemently on both sides.

In February the Washington State Liquor Control Board rejected our petition. We are not done with them. One member is sympathetic to our case, and one of the members against us has just been replaced by the governor. In the meantime the effort has drawn tremendous media attention both on the national and local levels. Interestingly, local interest heightened considerably after coverage by the New York Times, Wall Street Journal, Ad Week, and NBC News. My reading is that the public is very much on our side. I have received more approving letters and phone calls on this issue, than on any one in which I have been involved. The Medical Association has received many accolades for "hanging tough" against heavy odds.

In the meantime, the obnoxious beer ads continue unabated. I cannot imagine that the companies will ever voluntarily back off of their "lifestyle" themes. That is because 80 percent of beer is consumed by people in the 18- to 34-year-old age group. Shareholders will continue to look for the bottom-line, and go after "entry-level consumers," as they call them.

## Discussion

MR. FERGUSON: Abe, did the Wine Institute support you on this?

DR. BERGMAN: No, the California Wine Institute, whose advertising code we are promoting, said nothing. Even though our proposals would not affect them, the Washington wineries fought us. They are nervous about any advertising restrictions.

MR. TERET: What arguments have your opponents chosen to rely on?

DR. BERGMAN: Two main arguments. First, free speech. This is trumpeted by the media folks and the American Civil Liberties Union. The other main argument is that there are no conclusive studies to show that advertising is related to overconsumption.

MS. McLOUGHLIN: What are the main differences between advertising cigarettes and alcohol?

DR. BERGMAN: Well, Philip Morris owns Miller brewery, and interestingly Philip Morris has made a big thing about not targeting youth for smoking ads. One difference is the 21st Amendment that allows states to regulate alcohol commerce. Another difference is the history of television advertising of tobacco. It should be recalled that the removal of cigarette ads was a victory for the tobacco companies. That is because by giving up TV advertising, the cigarette companies also got rid of the counter-commercials that TV stations were obliged to show under the "fairness doctrine". Counter-commercials were hurting their sales. My feeling is that if we would have access to counter-commercials on the effects of overconsumption, I would gladly forget about restrictions on advertising.

MR. McGUIRE: I don't know if counter-commercials are effective or not, but the liquor industry hates them, and that tells me something.

MR. FERGUSON: I was interested, Abe, about your earlier comment about warning labels not being effective. I was under the impression especially on the ones on drinking during pregnancy, that they were fairly effective.

MR. McGUIRE: I have a rule of thumb. If the alcohol industry supports it, I question whether the measure is effective. The alcohol industry supported warning labels. The reason they supported it, I think, is that once they mention pregnancy and driving cars or machinery etc., it excludes them from products liability suits. Or at least the warning provides a defense.

# The Politics Of Compulsory Motorcycle Helmet Legislation: The Successful Washington Experience

Roy A. Ferguson

It was no easy task to pass a mandatory motorcycle helmet bill during the 1990 session of the Washington State Legislature. It took real commitment by a number of people. The motorcyclists' lobbying organization mobilized fully against the bill as we anticipated it would. *This was not a one year effort.* It took four years to get it re-enacted in degrees, actually 14 if you go back to '76 when the original helmet legislation was repealed.

## Historical Background

When the federal government turned the question of motorcycle helmets more or less over to the states in the mid-70s, the Washington Legislature repealed its helmet bill. Since then, one strong organization of motorcycle helmet users, ABATE (American Brotherhood Against Totalitarian Enactments) has fought reinstatement of the motorcycle helmet bill. The group's lobbyist was continuously on the scene and was quite effective especially with the totalitarianism argument.

We legislators do react to the public. I became involved in this issue just a month before I started my first term in January, 1987. Dr. Abe Bergman from the Harborview Injury Prevention Center sent all legislators a report on the economic consequences of not having a compulsory motorcycle helmet law. I wasn't familiar with the topic because I didn't ride motorcycles, but it really made sense to me. I quizzed him a bit and got a little more information and decided to introduce this as my first bill.

Here I was, a green legislator who really didn't know much about the process. I was from the minority party in the House of Representatives, where Republicans were outnumbered about two to one. Unbeknownst to me, a longtime legislator (Rep. Ruth Fisher) from the majority party introduced the same bill, probably because of Dr. Bergman's letter as well.

When it came to committee, the chairman gave us both hearings, which turned out to be pretty identical. After the first hearing I realized that my bill wasn't going to go anywhere, and so I said, "Gosh, Ruth, you've been here a long time, I'll help you and I'll lobby to get your bill through." We got it through the House easily, but it failed in the Senate.

I decided not to give up. The following January I introduced it again, but only after I went to the previous bill's sponsor and asked if she was going to introduce it again. She said, "No, go ahead." It passed the House by a wide margin, but it was altered in the Senate to cover only riders 18 years and younger, rendering it virtually ineffective; when a person is charging down the highway, the troopers can't tell whether he's 18 or 17. As a result, they were not enthusiastic about enforcing it.

We decided to try again in 1989. I looked hard, but couldn't find a sponsor in the Senate, so I asked myself, "Do we want to do this thing again?" It doesn't make sense to do it unless there's a reasonable chance of succeeding. At that time I consciously chose not to push the bill, but I introduced it and I got 40 co-sponsors in the House just to serve notice that we were interested.

### A Well-Organized Success

By this time I was getting a little more sophisticated about legislation and figured out how to do a better job. I realized was that we really needed a strong group to pass this. We needed some credibility, we needed some strong efforts, some follow-through and follow-up. So the summer prior to the January 1990 session we strengthened our coalition and met to refine our agenda, get our people organized, line up some commitments and get fully prepared. Various individuals agreed to work on the legislators, and I agreed to do a lot of the coordination.

This time we had things in place before we started. Our sponsor in the Senate was the chairman of the Health Care Committee, which heard the bill in place of the previously unfriendly Transportation Committee. The bill was included in the Head Injury Prevention Act of 1990, which covered public education and traumatic brain injury, not just motorcycle helmets. Mandatory helmets became part of a bigger program and we gained significant support through that broadness.

Not only was the Senate sponsor part of this coalition early on, he was a member of the majority party in the Senate which had not been enthusiastic about the issue in the past. He convinced the leadership to let the bill go if it had the votes. While the top three leaders voted against it on the Senate floor, the bill passed by a good margin. It's vitally important that we as legislative advocates find ways to work through the process. If an opponent is the No. 1 person in either house, we have to find a way to make that person vote their personal conscience on it and let the bill go.

The action then switched to the House of Representatives where the bill had traditionally been successful. It would have been easy to sit back, but we kept the pressure on and took nothing for granted. That strategy, as it turned out, was correct. During the House committee action, ABATE made a tactical error with a bogus memo from a "Dr. Z" which ultimately embarrassed the group. The bill progressed onto the floor of the House of Representatives, where it hit some highly vocal opposition but passed with a better than two-thirds margin. Minor amendments in the House version were accepted by the Senate and the bill was on its way to the governor.

Although the governor was a vocal supporter of the bill, we coalition members wrote letters, made phone calls and continued the pressure until he signed. We wanted to cover every base possible, and I think it helped. The ABATE people came to the signing to see whether he actually signed it or not. I have never seen the opponents of any bill come to a governor's signing. I guess they figured at the very end maybe he'd change his mind.

**Deciding Factors**

A number of important factors helped our situation. First, we took a statewide a poll to measure the level of public support. This became very important because it showed an overwhelming majority, 79 percent, was on our side. Prior to the session I sent every member of the Legislature a copy of the poll and a letter pointing out that voters in every area in the state favored this bill.

Information pieces were created and distributed to legislators. The State Department of Health and our Head Injury Foundation coalition members created an outstanding informational piece which talked about the problem in laymen's terms. We had information compiled by our Traffic Safety Commission and the experiences of Oregon and some other states. I also used the federal model even though it has not yet become law. In addition, we kept a media file with articles and favorable editorials about previous efforts to pass a helmet bill. These clippings were used to inform legislators of wide support for the bill.

At every step along the way I sent the poll and other pertinent data to newspapers in the State along with requests to editorialize in favor of the 1990 bill. I did it after the Senate committee vote, after the final vote in the Senate, after the final committee vote in the House and after the final vote in the House. A number of those newspapers editorialized a second time in favor of the bill.

In the Head Injury Prevention Bill we included some funding for education. For years the ABATE people argued that "we don't need this bill, but with some money from you we can educate our people against this type of risk." I agree that we should have provided that kind of funding. We did try, unsuccessfully, a couple times. But education's not the total answer. That was my argument and our coalition finally convinced enough legislators of that. When the opposition comes at you with education as the whole thing, it's important to broaden the bill to cover a number of issues, including funding for education. Not only did that split some of them off, it also blunted their argument that education could solve this problem completely.

**Lessons Learned**

A few things are essential when you're trying to pass tough controversial legislation:
- Have a good plan, follow through, make sure you cover every base.
- Don't give up after the first year if it fails. Some things are worth doing and this happened to be one of them.
- Form a strong coalition that's willing to work very hard and efficiently on behalf of the cause.
- Line up key legislators in both houses who are in favor of the bill and willing to work for its passage.

I believe that the motorcycle helmet legislation was a positive experience for me and for the state of Washington. The physicians from Harborview showed me that protecting against motorcycle head injuries was the right thing to do. As a result I've gained a vision on all safety issues. Each legislator goes to the capitol with a number of things in mind. Some may have only one or two items. Some may only know that they want to serve the public. I went with a couple things in mind. My younger child was physically and mentally disabled so I was going to work on disability issues and I was going to work on jobs and economic development.

Now I've broadened my agenda and I have five issues, one of which is safety-related concerns. I'm working on all the safety-related issues in the Legislature now because of this experience. My vision would be to find a cadre of legislators throughout the country that would be willing to do that sort of thing. Perhaps this group could identify some of them. There are a number of safety organizations in the country. We could promote in a number of ways the agenda items that are important.

# The Almost Successful California Experience: What We and Others Can Learn From It

Elizabeth McLoughlin, ScD

We were not so fortunate in California! The following list of issues and activities is designed for use by people working for passage of state motorcycle helmet laws (rather than a referendum or initiative). The generic points are followed by explanations of how these points played out in California.

**Know your legislature and legislative process.**

Get a state roster, so you'll have at your fingertips the names, addresses, districts and party affiliations for each legislator and other state elected official. Know the membership and legislative analysts for legislative committees pertinent to your bill.

> California's Secretary of State publishes the *California Roster, 1990,* which provides information on state, county, city and township officials; state officials of the United States and and a directory of state services. This book does not have committee assignments, but this information is available from the legislature's offices.

Understand the process and timetable by which bills are introduced, assigned to committee, heard, voted on, passed and signed in your state.

> In California, bills are introduced at the beginning of each session in the chamber of the bill's author, assigned a number (1991 helmet bill is AB 7). The bill is assigned to an "issue" committee [helmet bill to Assembly Committee on Transportation], a hearing is scheduled and a vote taken where simple majority of the committee is required for passage; if majority support is not achieved, the bill is dead;

if passed, it is heard by the Assembly Ways and Means Committee then voted on by the whole chamber, where a simple majority is needed for passage. If passed, it is delivered to the other chamber (for the helmet bill, to the Senate) where it is heard by the corresponding "issue" committee and the Appropriations Committee. If passed by both committees, it goes to the full Senate. If the bills passed by each house are not identical, they go to conference committee where a final text is agreed upon and then voted upon by each chamber. If passed by both chambers, the bill goes to the governor for signature within 12 days. A signed bill becomes law; a vetoed bill does not. The effective date of the provisions of the law is established in the bill.

## Know the legislative history of motorcycle helmet laws in your state.

*Federal legislation:* The highway Safety Act of 1966 and its implementing regulations strongly encouraged the states to enact mandatory helmet use laws. The act stated that the Department of Transportation would withhold 10 percent of federal highway construction funds and all federal highway safety funds from states that were not implementing an approved highway safety program. Under the Department's implementing regulations, a state's highway safety program would not be approved unless its law provided that motorcycle operators and passengers wore approved safety helmets. By 1975, the District of Columbia and 47 states required all motorcyclists to use helmets.

In 1975, the Department of Transportation began investigating whether federal funds should be withheld from non-compliant states. However, passage by Congress of the Highway Safety Act of 1976 withdrew the Department's authority to withhold those funds from states that failed to adopt and enforce laws requiring motorcycle riders 18 years of age and older to wear safety helmets.

Between 1975 and 1983, 28 states either weakened or repealed their helmet use laws. The most "typical" change occurred in 1977 or 1978, when states modified their laws covering all riders to laws covering only those under 18 years of age.

*California's response to the Highway Safety Act of 1966:* California is the only state which did not pass a helmet law between 1966 and 1975. Its senators and representatives were very active in lobbying Congress to withdraw DOT's authority to withhold highway funds from non-compliant states. In 1985, a law requiring motorcycle passengers under 15.5 years to wear helmets went into effect.

*AB 36 (1987):* In 1987, Assemblyman Dick Floyd (D) introduced AB 36. It passed the Committee on Transportation, the Ways and Means Committee and the full Assembly (41 yes, 29 no, 10 not voting) but was defeated in the Senate Transportation Committee, three votes short of majority. It was later granted reconsideration as a two-year bill. The Senate Transportation Committee subsequently conducted an interim hearing on key subject areas and issues of AB 36 on Nov 9th, 1987. The full testimony of this hearing was published

*AB 36 (1988):* In 1988, AB 36 passed the Assembly Transportation Committee, Assembly Ways and Means Committee, full 80-member Assembly (43 yes, 27 no); Senate Transportation Committee, Senate Appropriations, full 50-member Senate (23 yes, 12 no). It was vetoed by Gov. George Deukmejian, who argued in his veto message that it was "unfair" to require "responsible, mature operators" to wear helmets, and called for a law covering those under 21, the "young relatively inexperienced operators who have frequently engaged in the consumption of alcoholic beverages." State death, injury and blood alcohol concentration data do not support his arguments.

*AB 8 (1989):* In December 1988, Assemblyman Floyd introduced essentially the same bill. After legislative passage it was again vetoed by Gov. Deukmejian whose veto message stated that "This bill makes no attempt to address the concerns stated in last year's veto message."

*AB 55 (1989-90):* During the 1989 session, Assemblywoman Bev Hansen (R) introduced a "comprehensive motorcycle safety bill" which included a requirement that all riders under age 21 wear helmets, essentially, the governor's bill. During the debate on AB 8, all reference to helmets was negotiated out of AB 55. In the 1990 session, given the 1989 veto of AB 8, the under-21-years-of-age helmet requirement was reinstated in AB 55. Advocates for a full-use law then had to argue against passage of AB 55. The pressure against the helmet provision finally led to its removal before the bill was passed and signed. So one is not only fighting for certain bills; sometimes it is necessary to generate pressure to oppose counter-productive measures.

**Select the bill's author carefully.**

The political power of any bill's author is a key element in how the bill is handled by the legislature. There are times when advocates have no control over authorship; at other times, advocates can seek out an author to carry a certain piece of legislation. An appropriate author has a strong power base within the legislature, experience with and support of public health measures and a good track record in having his/her bills passed and signed.

> In California, the person who initiated the current legislative activity around helmets is Mary Price, whose 18-year-old son was killed in a motorcycle crash in 1985 while not wearing a helmet. In response to her son's death, Ms. Price vowed to work for a helmet law. Totally without experience in the political arena, she went to Sacramento and knocked on legislators' doors.
>
> The only person to give her a hearing was Assemblyman Floyd, who subsequently authored the three full-use helmet bills. Not always aligned with public health advocates (he was named the California legislator of the year by the NRA) and known as an aggressive, sometimes abrasive lawmaker, Floyd has worked vigorously for passage of these bills. His appointment as chair of the Assembly's Government Operations Committee in 1989 increased his power base and enhanced the bill's chances for passage.

**Identify obstacles to the bill's passage.**

Know and understand your opponents, both in the field (Hell's Angels, ABATE, etc.) and in the legislature. Chairpersons of committees have a great deal of power in determining how a bill is handled by that committee (by assigning or not assigning a hearing date and time, by managing testimony and debate during the hearing and by influencing the votes of the members). A hostile chair is very difficult to circumvent. Washington advocates circumvented a difficult member in the Transportation

Committee by framing their helmet bill as a public health (head injury prevention) bill rather than a traffic safety bill.

Sometimes it is useful to send the bill through a number of committees, thus avoiding a floor debate which tends to be difficult to control. Carefully tailor the hearing testimony to educate legislators about your issue.

> In California, the chair of the Senate Committee on Transportation in 1987 was strongly opposed to the helmet bill, even though he had sponsored full-use helmet legislation in the early 1980s. AB 36 was defeated in that committee, although the chair did permit it to be reconsidered as a two-year bill, and held a special hearing in November 1987. He was replaced by another chair in 1988 who was very supportive and AB 36 passed through that committee that year.

**Build organizational and grass roots support, with particular attention to constituents of key legislators.**

Regardless of how strongly a legislator supports a bill, a majority of his/her colleagues in the legislature must vote for it in order for it to become law. Elected officials want evidence of constituent support; their office staffs keep a tally of mail and phone calls on an issue. Opposition to the helmet law is highly organized and able to generate sacks of mail against the bill. This tends to be "pre-fab" mail in quantities that overwhelm mail in favor of the helmet law.

We need to counter quantity (lots of "pre-fab" messages) with quality (letters on official letterheads signed by local opinion leaders, respected citizens). To facilitate this quality support, it is valuable to have an active injury control network in place. It can then be activated in a timely manner when hearings are to be held or pressure from constituents needs to be generated.

> In California, Assemblyman Floyd's office was skillful in generating resolutions of support for a helmet bill from such groups as: orthopedic surgeons, pediatricians, university women, public health associations, trauma societies, developmental disability boards, insurance

companies, auto clubs, trauma education, American College of Surgeons, emergency physicians, highway patrolmen, hospitals and health systems and residential resources.

The Trauma Foundation has generated a *California Injury Prevention Network Directory* which currently lists more than 385 people who have agreed to be publicly known as injury control advocates. The legislative districts (Senate and Assembly) of each participant are included in the directory, which is available to participants both in hard copy and computer file. The computer file can be searched by district to identify network constituents of key committee members. Direct communication with these network participants can keep them informed about the voting records and current attitudes of their own representatives regarding the helmet bill, and can encourage them to put on the heat at critical junctures in the legislative process.

**Use the emotional and symbolic elements inherent in the helmet issue to advantage: This is inextricably linked with media coverage.**

Certain social/legislative issues generate powerful emotional reactions (e.g., guns, abortion, capital punishment). Motorcycle helmet laws generate similar levels of passion. Motorcycle riders and champions of "personal freedom" whose chant is "let those who ride decide" can be met by counter slogans, such as "let those who pay have the final say" or by the impassioned pleas of parents and spouses whose family members have died in motorcycle crashes. Television loves theater, and nothing brings out the cameras like a Harley-Davidson roaring through the capitol when hearings on the bill take place. While scientific data has its place in the process, the campaign's real energy derives from deeply held emotional positions on both sides of the issue.

The image of the "easy rider" as symbolic of what is right about California is held by a portion of the population. The motorcycle is identified with a freedom of spirit, an urge to live on the edge, the harnessing and control of extraordinary speed and power. For many riders, the motorcycle is less a means of conveyance than it is an extension of self. The most visible center of support for this view is, not surprisingly, Southern California and, specifically, Hollywood.

Hollywood stars had an extraordinary role in the campaign for a helmet law in California. Television cameras covered Evel Knieval's pro-helmet testimony during the first round of hearings. While showing a videotape of many of his daredevil jumps, he proudly boasted that he had broken every bone in his body doing these jumps except for his skull, because he always wore a helmet. *Motorcyclist* magazine attributed the governor's 1988 veto to the lobbying efforts of Hollywood personalities, with major influence attributed to comedian Jay Leno (the permanent guest host of Johnny Carson's Tonight Show) and Fred Dryer of the TV show "Hunter."

Gary Busey was one of the Hollywood stars who participated in a benefit to raise money to defeat the helmet bill. His head injury, sustained when he struck his head against the pavement after skidding his Harley, coincided directly with Floyd's re-submission of his bill in early December 1988, after the first veto. This brought an extraordinary amount of media attention to the helmet issue (national coverage on 20/20, state and local, TV, radio and newspaper stories). While injury control advocates cannot rejoice over another's misfortunes, it is a recognized fact that media coverage of individual or mass disasters can generate public support for injury control issues.

**Learn the principles and practices of media advocacy and pursue them vigorously during the campaign.**

Journalists like stories that have human interest. For better or worse, motorcycle crashes and the debate over helmet laws generate a lot of reader interest. Encourage journalists to mention whether a person involved in a motorcycle crash was wearing a helmet when they describe the nature and extent of injuries. Political cartoons can be a powerful tool in conveying a message in a single image.

To develop and maintain the best possible media image:

- visit editorial boards
- develop press packets
- cultivate relationships with individual journalists
- capitalize on late-breaking news events
- produce and distribute op-ed pieces and letters to the editor
- monitor and counter opponents' tactics
- train advocates to be effective on TV and radio

> California journalists have been giving excellent coverage to the last four years of work on the helmet bill. In general, the regular press has been very supportive. Local radio talk shows seem to enjoy hosting debates on the merits of helmet laws. Political cartoons in the daily papers have been right on target. We have used supportive cartoons in our newsletters and slide shows. Spot news clips on motorcyclist deaths and injuries now routinely report helmet use.

**Know your facts thoroughly and accurately. This includes general facts about the issue and state and local data on crashes, deaths, injuries and current use rates.**

The efficacy of motorcycle helmets to prevent the majority of head injuries is very well-documented (more so than almost any other "controversial" injury control measure. It is relatively easy to counter all of the opponents' objections (except those rooted in "personal freedom" arguments) with findings from research. It is valuable to identify at least one local "technical assistant" who can generate information sheets, counter opponents' arguments and testify at hearings.

NHTSA provides state-by-state data of motorcycle deaths in its Fatal Accident Reporting System (FARS) reports. The 1989 report became available in the spring of 1991. NHTSA also sponsors a "19 city observational study" of seat belt, child restraint and helmet use (*Restraint System Use in 19 U.S. Cities 1989 Annual Report.* Washington, D.C. NHTSA, June 1990).

State Departments of Transportation should publish annual statistics on motorcycle crashes, deaths and injuries.

> The Trauma Foundation published a newsletter in 1987-88 which has served as a general resource for questions and

answers about motorcycle helmet laws. It provides highlights of key scientific studies and reference list to help readers locate the original articles and reports. A new edition became available in February, 1991.

**Utilize the authentic voice of survivors and family members who have been affected by motorcycle crashes in support of a helmet law.**

Personal stories are compelling. However, for these stories to actually advance the prevention of injuries, these spokespeople must be knowledgeable about the issue, with their arguments grounded in the principles of injury control.

The involvement of injury survivors with disabilities is a relatively new development and not without some inherent difficulty for injury control. It is unacceptable for prevention campaigns to "use" people with disabilities as "exhibits" or "visual aids." It is acceptable for people with injury-caused disabilities to draw on their life experiences to inform and give shape to their advocacy for a less hazardous environment.

In California, Mary Price single-handedly started the recent activity for a helmet law (joining other notable people who have turned their grief toward constructive action, such as: Candy Lightner with Mothers Against Drunk Driving; Marilyn Spivak, National Head Injury Foundation; Peter Shields and Sarah Brady, Handgun Control, Inc.). In addition to founding an organization called Californians for Safe Motorcycling, Mary found an author and took a year's leave of absence from her job to lobby full time for the helmet bill. She has worked tirelessly through two vetoes and is poised for round three.

Steve Lambert, who was severely injured in a motorcycle crash, coordinates Survivors for Injury Prevention. He has given newspaper interviews, done a "Speak Out" spot for TV and testified at hearings in Sacramento. He works to involve California's Independent Living Centers in support of the helmet bill.

**Estimate the potential cost savings if your state had a full-use helmet law.**

The U.S. Supreme Court upheld a lower court (Massachusetts) finding that motorcycle helmet laws are constitutional. Because motorcycle injuries drain state financial, medical and social service resources, states have the right to protect these resources by requiring motorcyclists to protect themselves from serious head injuries.

Many political decisions boil down to economic issues, and this is true for the helmet law. In recognition of this, NHTSA has developed a formula for use in estimating potential state cost savings.

**Acknowledge that politics has a life its own, essentially unrelated to the merits of any particular bill.**

While passage of the helmet bill may be extremely important to its advocates, it is one of thousands of pieces of legislation considered each session. Bills and issues come and go, but legislators remain (with the high rate of incumbency, the membership may change very little over the years). Each legislator has his/her own agenda and favored bills. Individual bills can become currency for exchange as legislators swap support for a colleague's bill in exchange for one of their own. Alliances and hostilities develop and affect who supports which bill. It's also important for us to understand the issue of party loyalty.

**Be prepared, on a regular basis, for the roller coaster of emotions that follows your bill's victories and defeats.**

(Editor's note: The 1991 Session of the California Legislature again passed a motorcycle helmet bill. On May 20th is was signed into law by Gov. Pete Wilson).

## Discussion

MR. FERGUSON: I agree with your acknowledgement that politics has a life of its own. I think the essence here is to try to develop in legislators a vision for safety issues. As you mentioned, each legislator has an agenda, his or her own personal interest, but I find that in many cases it's not as big as you think it is. There is room for one good overall issue for the future, which can be safety-related issues.

MS. TRACY: Roy Ferguson's remarks do not fully reflect the agony that he and other proponents went through in Washington State. In the ten years that I've been lobbying, I've only seen two bills where I thought things got real down and dirty and nasty and scary: the AIDS omnibus bill and this one.

As the bill was passing, State Patrol protection was required for some of the legislators, and I was urged not to go to my car alone. I think people need to be warned of that. They need to know how nasty the opposition can get. There were no injuries - I don't want to imply that there were - but there were a lot of threats.

We were not only fortunate to have two pit bulls working on the Washington bill (Rep. Ferguson and Dr. Bergman), but we were helped by other things happening at the same time. There was the head injury report from the Department of Health, plus, very importantly, the results of a two-and-a-half year study on the development of a trauma system. Perhaps what got Senator West on our side (where he historically had not been) was his desire to sponsor the trauma system bill. They fit together very conveniently. In addition, the Republicans were looking at all sorts of health prevention things so they wouldn't have to look at national health type packages. All kinds of convenient little mixes can aid a prevention bill. But I still think it goes back to the pit bull mentality of the bill's prime advocates.

MR. CARPENTER: Then you cannot forget the legislators when you get the bill passed. When they are up for re-election, you need to get your network activated. One of the things we did was to get the Advocates for Highway and Auto Safety - which was very helpful throughout the process, sending experts to testify, etc. - to create a national legislative leadership award. Judith Stone, their Executive Director, came in and presented the award at press conferences in on both sides of the state. Sen. West, the Senate sponsor, had that ceremony videotaped and used it in his campaign advertising. It was a positive way to say thank you.

MR. FERGUSON: That's a good point. In politics we obviously get people who work for us and against us on different issues. I personally - and I'm sure Senator West would say so too - made a ton of friends on this issue. These individuals and groups were very helpful to me in reelection. I think that's important.

DR. BERGMAN: As a life-long Democrat it is difficult for me to go around the state and publicly support Republicans, but in this instance I did. If these politicians are going to stick out their necks for us, when re-election time comes, we have to help them.

# Efforts to Prevent Burns From Hot Tap Water

Murray L. Katcher, M.D., Ph.D.

In May 1978, I was completing my pediatrics residency and had accepted a position on the faculty of the Department of Pediatrics at the University of Wisconsin-Madison. While attending the annual meeting of the Ambulatory Pediatric Association, I heard Ken Feldman present his excellent review of tap water scald burns in Seattle children (Feldman, et al., 1978). Like many others in the audience, I was very impressed by the severity of these burns as well as by the tremendous potential for prevention. Although the use of safety restraints requires a given action *each time*, the *one-time* lowering of the water heater thermostat to 120°F may prevent many tap water scald burns. During the following year I brought the message to residents and my faculty colleagues, and we incorporated educational counseling about tap water scald prevention into the anticipatory guidance portion of our health supervision visits.

At the University of Wisconsin Hospitals in 1979, almost all children with burns of less than 50 percent burn surface area were managed in the pediatric intensive care unit (rather than in an adult burn unit) with consultation from a burn surgeon. When the general pediatrician who cared for these children left the pediatric department one year later, I became the "pediatric burn doctor." At that time I became directly aware of a number of children with preventable burn injury, including those who had been burned from hot tap water. This paper traces our efforts in Wisconsin to gather data, educate the public, and create legislation that would result in the prevention of burns from hot tap water.

## Defining the Problem

In 1980, we began a ten-year retrospective review of the charts of all patients hospitalized in Dane County, Wis., as a result of tap water scalds (Katcher, 1981). Of 33 patients, we found that 88 percent had readily identifiable risk factors: 17 (52 percent) were children younger than 5 years; 3 (9 percent) were older than 65 years; 10 (30 percent) were physically or mentally disabled; and another patient was burned in a non-home environment. Of five deaths, three occurred in children younger than 30 months, and two occurred in patients older than 70 years. An annual rate of 1.05 hospitalizations per 100,000 persons was found.

The *problem* is that (1) many people are unaware of the danger of burns from hot tap water; (2) water heater thermostats in many households are set at dangerously high temperatures; and (3) when exposed to these temperatures, even for a brief time, high-risk individuals may suffer full-thickness burns, which may result in disfigurement or death. The *solution* is to lower the maximum household water temperature to 120-130° or less, which would prevent most of the severe burns.

A review of time and temperature data shows that exposure to water of 140-150°F will cause a full-thickness burn in 2-5 seconds in adults (Moritz and Henriques, 1947) and even faster in children (Feldman, 1983); however, a full-thickness burn requires approximately 30 seconds of exposure at 130°F and 10 minutes exposure at 120°F. At these lower temperatures, usually adults or children can remove themselves (or be removed) before severe burns occur. Because some children may receive severe burns at approximately 130°F, we recommend 120°F as an optimal safe temperature; most household chores requiring hot water can be readily performed at this lower temperature.

## Prevention Campaign
### Multimedia Public Effort

In early 1981 a committee consisting of members of local utilities companies, the medical profession, senior citizen groups, and pharmacists was brought together by the Madison Department of Public Health in order to begin a tap water scald prevention campaign. The utility company included an information supplement with its monthly bills and offered to measure and adjust hot water temperatures. Educational pamphlets were distributed to primary care practitioners and to senior citizens groups. Pediatricians were urged to include tap water scald prevention in anticipatory guidance messages to parents of young children. In addition, the Mayor of Madison declared "Tap Water Scald Prevention Week," during which time public service announcements were aired with prevention messages on television and radio.

At that same time the Wisconsin Assembly Commerce and Consumer Affairs Committee also held an informational hearing about the danger of hot tap water and the energy savings that could be attained by lowering the water heater thermostat. Although testimony from several sources was given to the Committee, no legislative action was taken. However, the hearing did generate statewide publicity about this danger.

Shortly after the above ten-year retrospective study was published (Katcher, 1981), I was contacted by the Wisconsin Electric Power Company to aid in an energy-conservation campaign that involved the lowering of water heater thermostats. I agreed on the condition that I be allowed to evaluate the effectiveness of the program (Katcher, 1987). Wisconsin Electric provides power to a service area of 2 million people in and around Milwaukee, Wis. When faced with the possibility of needing an additional power plant to meet increased consumer demand, the company became interested in the conservation potential of lowering water-heater energy consumption.

During August 1982, a one-month multimedia tap water scald prevention program was conducted. A pamphlet was enclosed with the electric bills sent to the company's approximately 750,000 residential subscribers. Subscribers were invited to request a free liquid-crystal thermometer for tap water testing and an educational brochure. The brochure described the danger of hot tap water and explained both how to use the thermometer to measure the maximum water temperature at the tap and how to lower the water-heater thermostat setting, if necessary. Emphasis was also placed on the energy savings that could be attained from the lower setting. During the same month educational messages about hot tap water appeared in all major newspapers in the greater Milwaukee metropolitan and outlying areas and on prime-time television and radio. These messages provided a toll-free number for requesting the free educational brochure and thermometer. In addition, doctors' offices, hospitals, and social service agencies received posters and pamphlets containing postage-paid coupons for ordering the free materials. More than 140,000 requests were received.

Pre-program and post-program random surveys of the general population found increased awareness of the danger of hot tap water, from 72 to 89 percent, after the campaign, but no increase in testing or lowering of water-heater temperatures. A third random sample survey among those requesting thermometers found a higher rate of testing (61.5 percent) as compared with the general population (3.4 percent) ($p<0.001$). Of those reporting high temperatures, 52 percent lowered their water-heater thermostats. As a result of this effort, thermostats of an estimated 20,000 water heaters were lowered from dangerously high levels.

From this study we concluded that a large proportion of people (28 percent) were unaware of the danger of hot tap water. In addition, household experience with tap water burns was greater than previously appreciated. (Of those surveyed, 11.8 percent reported someone in their household had been previously burned by hot tap water.) A multimedia educational campaign may increase public awareness but not necessarily change behavior. However, among those persons motivated to send for a free thermometer, many will test their water-heater temperature, and if it is found to be high, more than half will lower the thermostat.

**Anticipatory Guidance**

Because of the special hazards to children, the American Academy of Pediatrics suggests that physicians include tap water safety discussion as part of an injury prevention anticipatory guidance program in the office. Gregory L. Landry, M.D. (then a pediatric fellow) and I looked at two different methods of office counseling (Katcher, et al., 1989). Pediatric clinic clients were randomized into two groups: (1) those receiving an informational pamphlet, a one-minute discussion about tap water safety, and a baseline questionnaire; and (2) those receiving these interventions plus a liquid-crystal thermometer for testing maximum hot water temperature.

One month later in a follow-up interview, the impact of the added thermometer was measured on (1) knowledge regarding scalding, (2) hot water temperature testing, and (3) water heater thermostat lowering. We found that the group receiving the thermometer was more likely to test the water temperature than the "pamphlet only" group (46.4 percent versus 23 percent; $p<0.001$). Within the "thermometer" group, reported testing was highest (59.1%) when the counseling was provided at a well-child visit for a child younger than three years and lowest (34.2%) following a younger-child sick visit ($p<0.001$) (Shapiro and Katcher, 1989). In households where the reported water temperature exceeded 130°F, 77.3 percent reported lowering the thermostat setting, independent of the study group. Our conclusion was that the use of a relevant facilitating device (such as a liquid-crystal thermometer) in office anticipatory guidance efforts may increase behavioral compliance.

## Postpartum Counseling

Mary Melvin Shapiro, a medical sociologist, and I then looked at injury prevention education during postpartum hospitalization, as it relates to tap water scald prevention (Shapiro and Katcher, 1987). One might think that injury prevention education in the postpartum period may not be effective because the new mother receives much information relating to the care of her new baby and herself. Also family stress often accompanies the birth of a baby.

In the packet of information given to all women at the time of admission to the maternity ward of three hospitals, we provided a pamphlet about the danger of scalds from hot tap water and a free thermometer for testing hot water temperature. Women were randomly assigned to two groups: (1) those who received the pamphlet and thermometer alone as part of the admission packet (control group), and (2) those who received, in addition, a one-minute explanation about the danger of hot tap water, calling to their attention the material in the packet (explanation group).

Several months after the perinatal hospitalization, 604 mothers were reached by a follow-up telephone survey. There was a higher degree of tap water burn awareness and of temperature testing (both $p<0.001$) in the group that had received the brief explanation in addition to the printed material and thermometer, as compared with the group that received no explanation. If dangerous temperatures were found, individuals were equally likely to lower the water-heater thermostat, regardless of the group.

We concluded that perinatal nurses may enhance patient education by personally reviewing selected portions of written patient education material with new mothers prior to discharge from their maternity hospitalization.

## Legislative Efforts

While all these patient, provider, and community education efforts were taking place, the American Academy of Pediatrics had developed model tap water scald burn prevention legislation patterned after bills passed in the states of Washington and Florida. I thought it would be easy to have a similar law passed in a progressive state such as Wisconsin, especially since there was no fiscal note attached. The State Medical Society lobbyist agreed and found several sponsors. Model bills from other states were given to the Legislative Reference Bureau for drafting. I then left the bill in the hands of the Medical Society lobbyist, who had a very full bill agenda of his own. Unfortunately, without close attention, the bill was never written up and never introduced to the legislature.

During the next session I became more proactive; and with the strong support of several organizations, including the State Medical Society, Wisconsin Chapter of the American Academy of Pediatrics, Wisconsin Public Health Association, and the Maternal and Child Health Coalition (a strong advocacy organization consisting of representatives of more than 40 organizations), a four-point bill was drafted and introduced. The contents of the bill are summarized as follows:

- All new water heaters sold in Wisconsin shall be preset at the factory at a temperature of 125°F or less;
- A label warning of the danger and unnecessarily high energy consumption of a setting above 125°F must be plainly visible;
- All public utilities furnishing gas or electricity must place a warning in the billing statement at least annually; and
- A landlord of a dwelling where a water heater serves a single unit must set the water heater thermostat at 125°F or less before occupancy of each new tenant.

An interested member of the Legislative Counsel made suggestions about which committees would be most likely to approve the bill, once introduced. The bill started in the Assembly Commerce and Consumer Affairs Committee. A letter-writing campaign was organized to reach each committee member. After a public hearing the Committee approved the legislation 10-0. The bill then proceeded to the Assembly floor, where, since all nine sponsors were Democrats, the bill was perceived to be a partisan effort, and a heated debate ensued. One of the sponsors argued "…and what if I must leave my child [in the bath tub] to answer the doorbell," at which point a member of the opposition party said he should be considered "guilty of child abuse." The argument continued and the vote was postponed. During the interim, influential Republicans were lobbied; the bill was reconsidered several days later and was passed by the Assembly 69-25.

Several months later, when the bill went to the Senate Judiciary and Commerce Affairs Committee, a national group, the Gas Appliance Manufacturers Association (GAMA), sent a representative to oppose the bill at a public hearing. The gist of his argument was that unfriendly actions toward the industry would result in higher water heater prices to Wisconsin consumers (as a result of the "special" thermostat setting).

Fortunately, the Committee approved the bill with a 5-2 vote. The bill was then to move to the Senate floor.

Prior to the Senate vote, two main forces of opposition emerged. First, a strong letter of opposition was written to the major Senate sponsor by one of his constituents, the mayor of the largest town in his district. (The mayor happened to be a plumber.) The heated letter stated that the bill was "bad for plumbers" and that he would actively work to unseat the Senator should the Senator continue to support the bill. The State Medical Society lobbyist arranged a meeting for the Senator with the plumber/mayor and me, at which time we convinced the constituent that the bill would not affect plumbers. After seeing some pictures of burned children, the constituent was convinced. He stated that his family had been in the plumbing business for several generations and that he had never heard of a burn from hot tap water. My reply was that when a child gets burned, the parents usually don't call a plumber.

The second Senate "blockade" came from an eloquent and influential Senator whose election opponent had gone on to be elected to the Assembly, where he subsequently became the major Assembly sponsor of the bill. The Senator did not want to support any bill initiated by that Assemblyman. In fact, the Senator wrote to his constituency suggesting that the bill would force individuals to have certain water heater settings in their homes. Subsequently the Senator received complaints from water heater salespeople and other constituents. Once again a personal meeting convinced the Senator that the bill would only affect (out-of-state) manufacturers of water heaters. By the end of the meeting, he reluctantly agreed not to speak out against the bill on the floor of the Senate. His wife, a pediatric nurse practitioner, later influenced him to support the bill, and he voted for it on the Senate floor. The bill passed and was on its way to the Governor.

The Governor, believing that the hot water safety bill was anti-business, was planning not to sign it. However, he was influenced by several of his close friends, who were urged by advocates to call and write him. In addition, there was a letter-writing campaign to the Governor, which resulted in more than 100 letters pertaining to this fairly innocuous bill. The governor signed the bill on Nov. 27th, 1987, and it became "1987 Wisconsin Act 102." When he signed it the Governor stated that letters from many citizens and groups had influenced his natural inclination not to sign the bill.

### Industry Guidelines

Several months earlier, when the GAMA representative was in Madison for the Senate public hearing, I mentioned to him that Wisconsin was just one of many states that would eventually have similar legislation. The legislative effort was being coordinated by the American Academy of Pediatrics, and pediatricians throughout the country were supporting it at the state level. However, each state law would be slightly different in the temperature settings and labeling requirements, and that variability would cause much difficulty for the water heater manufacturing industry. I tried to convince him that it would be easier for manufacturers to recommend an industry-wide change in water heater settings (self-regulation). I also mentioned the potential liability of the water heater manufacturers for burns resulting from high preset temperatures.

Several days after the Governor signed the Wisconsin bill, the GAMA legal counsel asked me to participate in a national meeting of all interested organizations to discuss uniform settings and labeling instructions. He hoped to stop the proliferation of state laws, each of which would have different requirements for the water heater manufacturers. In March 1988, almost ten years after the Feldman report, a meeting occurred at GAMA headquarters in a suburb of Washington, D.C. The meeting included representatives from the American Academy of Pediatrics, the Consumer Federation of America, and the National Safety Council. We agreed that (1) all water heaters should be preset at a safe temperature, 120-125°F; (2) new labels detailing the dangers of hot water burns should be required; and (3) installation manuals should be changed to encourage the installer to leave the water heater at the lower (preset) temperature.

During the next few months GAMA, working with all U.S. manufacturers of gas and electric water heaters, fully cooperated with this effort. As of now, all new electric water heaters have appropriate labeling and instructions and are preset at 120°F. (Underwriters Laboratory has not changed its recommendation of 140°F; however, all manufacturers are using the voluntary 120°F setting.) Gas water heaters are set in the "off" position; however, the "detent" position (the thermostat setting used by the installer) is now at 120°F.

### Conclusion

Although new water heaters are likely to be preset and installed at safe temperatures, the average "lifetime" of a water heater is approximately ten years; therefore, on the average, only 10 percent of the nation's water heaters will be replaced each year. In the meantime it is essential to continue to educate individuals about the danger of hot tap water burns, the importance of testing the water heater temperature, and the importance of lowering the water heater thermostat if the hot water temperature exceeds 130°F.

There is much work to be done in the field of injury prevention, and all the components (education, technologic interventions, and legislation/regulation) must be coupled with strong advocacy in order to prevent injuries and deaths in young children.

The coalition that was built around this injury prevention legislation, 1987 Wisconsin Act 102, has continued to function and had a major influence on the passage of legislation establishing a permanent, state-funded injury prevention unit in the Wisconsin Division of Health. But that is a another story!

## References

Feldman KW. Help needed on hot water burns. *Pediatrics* 1983; 71:145-146.

Feldman KW, Shaller RT, Feldman JA, et. al. Tap water scald burns in children. *Pediatrics* 1978; 62:1-7.

Katcher ML. Prevention of tap water scald burns: Evaluation of a multimedia injury control program. *Am J Public Health* 1987; 77:1195-1197.

Katcher ML. Scald burns from hot tap water. *JAMA* 1981; 246:1219-1222.

Katcher ML, Landry GL, Shapiro MM. Liquid-crystal thermometer use in pediatric office counseling about tap water burn prevention. *Pediatrics* 1989; 83:766-771.

Moritz AR, Henriques FC Jr. Studies of thermal injury: II. The relative importance of time and surface temperature in the causation of cutaneous burns. *Am J Pathol* 1947; 23:695-720.

Shaprio MM, Katcher ML. Targeting pediatric injury counseling: A case study in tap water burn prevention. Presented before the American Public Health Association, 117th Annual Meeting, Chicago, Ill. October 25, 1989.

Shapiro MM, Katcher ML. Injury prevention education during postpartum hospitalization. Presented before the Ambulatory Pediatric Association, 27th Annual Meeting, Anaheim, Calif., April 30, 1987. *Am J Dis Child* 1987; 141:382.

## Discussion

MR. McGUIRE: When are antiscald devices in showerheads going to be mandated?

DR. KATCHER: Antiscald devices are another effective way of preventing burns, and through building code ordinances one could get those into all new buildings. This technology and regulatory approach is being strongly and successfully supported by the National Safe Kids Campaign. I believe the Wisconsin building code requires antiscald devices in bathroom shower heads in new buildings. Antiscald devices won't prevent burns from all household faucets, nor will turning down the water heater temperature once. People can raise the thermostat to higher temperatures. However, we have made home visits to people, some of who said they had turned it down; over all we found about 70 percent agreement with their reporting.

DR. WILLIAMS: What has been the effect of the 1988 law?

DR. KATCHER: Dr. Rivara, would you like to comment on the Seattle follow-up study?

DR. RIVARA: Dr. Ken Feldman and I did a survey of Seattle-area homes that had water heaters installed after the new law went into effect in 1983. In 85 percent of those homes the water heater remained at the preset temperature. However, we also had a control group of homes that did not have new water heaters installed in that time period, and about 70 percent of those were also at the lower temperature. In Seattle it was a combination of the law and the education.

DR. BERGMAN: When you say "set the temperature," did you measure the water coming out or did you look at what it was set at at the hot water heater.

DR. RIVARA: Both.

MR. McGUIRE: That's another interesting question. Is there a correlation between what it says on the hot water heater and what comes out three rooms away?

DR. KATCHER: Sometimes there is no correlation. We talked about testing it at the tap because we have had experiences where the temperature of the water was 170°F and the thermostat is set at 140°F, and sometimes in the reverse direction. At least one study has shown little temperature drop (<1°C) with distance from the water heater to the faucet.

# The Case of The Fire Safe Cigarette: The Synergism Between State and Federal Legislation

Andrew McGuire

The idea of altering cigarettes to prevent them from starting fires has been studied for more than 100 years. Several efforts to encourage or mandate fire safety in cigarette design have been mounted. Nevertheless, only in the past 13 years has a solution appeared near. During that time a campaign involving the media, personal contact with opinion leaders, the civil justice system and, especially, combined state and federal legislative initiatives has been waged against the apparently limitless resources and intransigence of the tobacco industry.

A look at the current campaign for the fire safe cigarette reveals some strategic insights about accomplishing change through legislation. With the signing of the Fire Safe Cigarette Act on Aug. 10th, 1990, President George Bush approved funding for the Center for Fire Research to develop a fire safety standard for cigarettes. As a result, cigarettes sold in America will be fire safe sometime in the mid-1990s.

## The Cigarette Fire Problem in America

According to the U.S. Fire Administration, 1665 people died in 1988 as a result of cigarette fires in the United States. In addition, more than 4300 injuries and $443 million in property damage resulted from the 230,000-plus fires attributed to smoking materials. Cigarette-caused fires are the leading cause of fire death in the U.S. and account for 25 percent of all fire fatalities.

Past attempts to prevent these tragedies have included failed and flawed public safety campaigns with exhortations like "Be careful when you smoke" and "Don't smoke in bed." Other efforts have focused on state and federal flammability standards for upholstered furniture and mattresses. This long-term solution doesn't affect existing products, which have an average life span of 30 or more years. Cigarettes, on the other hand, are in the stream of commerce for approximately three to four months. Thus the current campaign, which began in 1979, seeks to prevent cigarette-caused fires by mandating fire safety standards for cigarettes. Once these standards are implemented, fire deaths and injuries caused by cigarettes will be significantly reduced in a matter of months.

## The Fire Safe Cigarette: 1854-1979

This is not to say the idea of creating a cigarette that would not set fires suddenly sprang into being in 1979. The first patent for a "self-extinguishing" cigarette was issued on Aug. 1st, 1854 (Bristol #11,409). At least 120 U.S. and foreign patents succeeded it, including schemes to treat cigarette paper and tobacco with heat and ash-retarding chemicals; to use foil, fiberglass or asbestos as a cigarette sleeve; to add physical devices to extinguish cigarettes at a certain point, and to modify cigarette holders.

More recent solutions have included a combination of the following: altering the circumference of cigarettes (slim cigarettes are less prone to start fires), using "puffed" tobacco (already used in low tar/nicotine cigarettes), changing the porosity of cigarette paper, using filters and other subtle design modifications of existing cigarette-making techniques. By 1981 it was clear from research at the National Bureau of Standards that a cigarette did not even have to self-extinguish to reduce ignition propensity. Nevertheless, the industry has steadfastly refused to test-market any fire safe product except for R.J. Reynolds' "Premier" which failed in the test-marketing phase and was abandoned.

There have been at least three prior attempts to raise the issue in the political and regulatory arena. In 1929 Congresswoman Edith Norse Rogers from Lowell, Mass., successfully pressed the National Bureau of Standards to conduct research into the feasibility of making what she called a self-extinguishing cigarette. In an AP story, the Boston Herald American (March 31st, 1932) reported that the bureau had proved "the practicability of a self-extinguishing cigarette..." The final sentence quotes Dr. Lyman Briggs, acting director of the bureau as saying: "All there is to do is find a manufacturer to take up the idea." No manufacturer "took up the idea." However, in 1939 advertisements featured "slower-burning Camels" that were less likely to start fires.

In 'Readers Digest' (September 1950) a article titled "So You Want To Burn To Death" devoted about three paragraphs to the fact that the California state fire marshal and the National Bureau of Standards had come up with a way to make cigarettes self-extinguishing. However, it observed:

"The tobacco industry is not yet ready to make its product less of a fire hazard. The faster cigarettes burn, the more are used, the bigger the sales."

There is no record of a resulting campaign, though the Bureau of Standards and California state agencies such as the Fire Marshall's office and the Bureau of Home Furnishings continued to study cigarette ignition properties fitfully in the ensuing years. The tobacco lobby, if it wasn't powerful in 1931 was getting more powerful in 1951. It was especially powerful in 1971.

When the Consumer Product Safety Commission (CPSC) was created in 1973, tobacco industry lobbying insured that its enabling legislation deprived it of any direct authority over tobacco products. However, development of the CPSC and, prior to that, the Flammable Fabrics Act Amendments in '67 did allow the federal Center for Fire Research to start research on eliminating the fire hazard caused by cigarettes. Because they had no jurisdiction over cigarettes, they tried to regulate the flammability of fabrics, furniture, mattresses, carpets and, as we have heard, sleep-wear for children.

At that point, the furniture industry got a little upset and said, "Wait a minute. Furniture does not spontaneously ignite. Why regulate furniture? Why not the ignition source, the cigarette?" The furniture industry revived the idea of a "self-extinguishing" cigarette by getting the late Sen. Phil Hart of Michigan to introduce in a bill in 1974 mandating that America-produced cigarettes be self-extinguishing within 10 minutes. The scientific justification for that was incredibly shoddy. Looking back, it probably would have prevented about 2 to 8 percent of cigarette fires. The tobacco industry was very upset that it snuck through the Senate and, despite strong support from consumer advocate Rep. John Moss, it was easily killed in his House Commerce Committee.

### Launching the Fire Safe Cigarette Campaign

In the late 1970s, as executive director of the Burn Council at San Francisco General Hospital, I had been cultivating relations among firefighters and burn surgeons for several years. I kept hearing people saying we should go after the cigarette, and by 1978 I decided to get serious about the issue. I took my file, then about a quarter inch thick, to some investigative journalists in San Francisco and helped them obtain funding from the Oakland fire-fighters local to interview people at the Center for Fire Research, the Tobacco Institute, Sen. Phil Hart's office and elsewhere. The result was an article published in 'Mother Jones' (July

1979) stating that it was possible to make a fire safe cigarette, but due to the power of the tobacco lobby it was politically impossible to make it happen.

In May 1979, I organized simultaneous press conferences by fire chiefs, burn care specialists and others in 14 American cities to announce a campaign for fire safe cigarettes. Each spokesman had my press release, a copy of the 'Mother Jones' article, and, extremely importantly, videotapes for use on television news. The video showed a one-minute time-lapse image of a cigarette igniting a chair over a three-hour period.

Those press conferences were triggered by endorsements of the fire safe cigarette idea, first by the American Burn Association and, shortly thereafter, by the International Association of Fire Chiefs. Those were the first two organizations I approached. Not for two and a half years did I achieve my goal of getting an endorsement from the American Medical Association. After about 45 or 50 national organizations had endorsed, the AMA in a sense got shamed into it. At the same House of Delegates meeting, it also got rid of the Philip Morris shares in the AMA trust fund.

In October 1979, Rep. Joe Moakley from Boston introduced a bill giving the Consumer Products Safety Commission jurisdiction over cigarettes as a fire hazard and the power to ban cigarettes unless they stop causing fires. By that time, the public information generated by the press conferences and the 'Mother Jones' article was beginning to spread. California Sen. Alan Cranston introduced a companion bill into the Senate in January 1980. In 1983, Sen. John Heinz joined Cranston as a co-sponsor.

My strategy was to push for simultaneous federal and state bills mandating that cigarettes not cause ignition. That was a viable idea because there was absolutely no federal law with anything to do with cigarettes and fires. In other words, states could, and still can, act without preempting any federal law. I knew the tobacco companies would find federal legislation more palatable than the prospect of facing 50 different state standards. Furthermore, if scientifically respectable standards were produced, published and adopted anywhere, they could be expose the industry to enormous liability in cigarette fire lawsuits.

## The Legislative Campaign in the States

State bills, using almost the same language, were introduced in 13 states over the past 11 years, beginning with California in April 1979 and concluding with Wisconsin in 1989. I spent three years in the early 1980s working those state legislatures and cooperating with a network of approximately 500 people around the country. Most of these were fire chiefs, fire marshals, burn doctors and nurses, plus a few consumer activists or survivors whose children had died in cigarette fires.

It sounds ghoulish, but to get attention in the state legislatures I tied my efforts into cigarette-caused fires that killed children. A number of those states already had activity resulting from multiple deaths of children in fires. My method was go to the fire prevention bureau, which is always in close contact with the local media, and coach the local fire marshal say, "This could have been prevented had been federal bill in place." Or "This could be prevented if the state bill goes through the legislature." This wasn't always necessary; enlightened people in some legislatures simply took it on as a good public health issue. But for the most part state legislative involvement was triggered by multiple fire deaths of children.

I consulted on all successive versions of bills and strategy decisions to avoid co-option. The industry made vigorous attempts to shape opinion among firefighters through grants to fire departments for fire prevention. I also informed state legislators of the part their efforts were playing in the federal legislation process.

The fire safe cigarette was always described as a fire-prevention, burn prevention issue, never an anti-smoking issue. That was incredibly important from the beginning. In fact, for the first five years I fended off any anti-smoking folks who wanted to get involved. Although I was totally sympathetic with that issue, I didn't want it to dilute or taint or bias what this effort was all about.

The first federal hearing (in 1983 in Henry Waxman's House subcommittee) was triggered by all of the activity in the states. The tobacco industry went along with the hearing because it figured "we'll just bring this up on the federal level, and get something done that will pre-empt all the state efforts." It assumed that whatever happened on the federal level would then go away because the industry had a lot of control in '83.

Despite the hearings, it was clear that the tobacco lobby had enough power to stop anything from happening. That's when I increased my state lobbying efforts, especially in California and New York. In New York we almost succeeded. Our bill passed in the Assembly by about 126 to 18. We got it into the Senate with 31 co-sponsors out of 60 senators. It went through the policy committee and passed, and went on to the finance committee where the chair was a co-author of the bill. Governor Cuomo had held a press conference a week earlier saying he was ready to sign it.

Then we learned another of those lessons about the unpredictability of politics. The finance committee chair was hospitalized and the vice chair killed the bill by not letting it out of the committee.

This near-win in New York was decisive in gaining passage of the federal Fire Safe Cigarette Act of 1984. It persuaded the Tobacco Institute to agree with Moakley, Cranston and Heinz on legislation authorizing a three-year study of the technical and economic feasibility of a fire safe cigarette and appropriating $3 million to the National Bureau of Standards (now the National Institute of Standards and Technology) to conduct the research. This answered the industry's main fears that immediate passage of state laws would provide insufficient time to design a commercially viable fire safe cigarette and would allow trial lawyers to pursue product liability claims on the basis of a promulgated state standard.

### Federal Legislation and Summary

President Reagan signed the Cigarette Safety Act on Oct. 30, 1984. Three years later the results were delivered to Congress: yes, a fire safe cigarette is feasible with very minor modifications in construction. No chemicals are added or subtracted. The paper is a little less porous, the tobacco is puffed up to decrease its density, and there must be a filter tip.

Using the ingredients at their factories, the cigarette companies had made prototype cigarettes with these physical characteristics. They pumped them off their regular machines at 8,000 a minute. Fire safe cigarettes looked the same, they smelled the same, they smoked the same. At the Center for Fire Research, the prototypes were placed on flammable furniture for small scale and full scale testing. They ignited no furniture. When researchers lit one of these cigarettes and put it on furniture, it would burn down to the filter tip, but without enough heat transfer to cause ignition. Why? We don't know. The Center's fire physicists want to investigate that.

Spurred on by the good news, Moakley, Cranston and Heinz introduced the Fire Safe Cigarette Act in the House and Senate in late 1987. Again the Tobacco Institute exerted its political muscle, and the bill was stalled in the House. Again I went to state legislators for help. State bills were introduced in Minnesota, California, Massachusetts, Wisconsin and New York and, once more, federal legislation was accepted by the tobacco industry. This act requires development of a federal fire safety standard for cigarettes by 1993. If, at that time, the cigarette companies do not comply with the standard, I will ask as many states as possible to introduce bills based on the federal standard. They'd be saying, in effect, "we don't trust that the tobacco industry is going to let this happen on the federal level." As long as the state level folks understand that they are playing a role in a much larger picture, that pressure is very helpful.

However, according to "reliable rumors," the cigarette companies are so fearful of product liability suits that they will welcome federal regulation. It is predicted that final federal regulatory law will be enacted quickly after the standard is promulgated. At that point, the states will have done their job, and fires caused by cigarettes will become a horror of the past.

## Discussion

DR. BERGMAN: Why won't companies make fire safe cigarettes on their own? Are they afraid of lawsuits?

MR. McGUIRE: It's a Catch-22. The longer they hold this off before any lawsuits occur, the less money lost through lawsuits.

More importantly, they want to minimize the amount of negative publicity. As we've seen with the Ford Pinto and other cases, once these things start, there's a rash of out-of-court or in-court settlements. Until recently, the tobacco industry's statement has been: "it's technically, scientifically impossible to make a fire safe cigarette. It's a great notion, we'd love to do it, we've been working on it for years but it can't be done". Once the federal study came back saying yes, it can be done, they started saying it's not commercially feasible. They claim (and it's a sham) that until they test-market any modification, they don't know whether people will purchase the cigarette. That doesn't sell in Congress or in a court room. They are about to meet their maker in the courts.

DR. KATCHER: Could the industry get together and say there may be liability and we'll all go with this?

MR. McGUIRE: They won't get together because, first of all, they hate each other. Secondly, that would raise anti-trust issues. I have evidence, from a New Jersey court case, that Philip Morris has done a lot of research on a fire safe cigarette that would look, smell, taste and be like a Marlboro. That research program was started in 1981.

MR. MARTIN: Can you expound on the industry's costs for making that cigarette? Do you have any economic cost estimates?

MR. McGUIRE: When R.J. Reynolds and Lorilar joined forced to make the prototypes, the actual cost was estimated at 3 percent cheaper than normal. They use less tobacco.

MR. FERGUSON: Could we at the state level look at a bill that would agree with the concept now instead of waiting two years for the standard to be developed?

MR. McGUIRE: Wait. In '84 when we negotiated the language for the first bill, which was signed by Reagan, I made a commitment (although I really have no power to do this) that I would call off all the state bills. I have made the same commitment to the current process. I've promised that the cigarette industry will not be hassled, at least by my efforts, for two years.

DR. BERGMAN: I an struck by the linkages you forged with other advocates such as the fire fighters, a politically popular and powerful group.

MR. McGUIRE: You are absolutely right. The Tobacco Institute identified the fire service as its major threat; not me, nor the burn doctors. In its attempt to shape opinion among fire fighters, the Tobacco Institute has provided fire services with an estimated $4 million in fire prevention grants between 1983 and the present.

A wonderful article by Myron Lavin in The Nation (July, 1989), entitled "Tobacco Smoke Screen, Fighting Fire with PR," documents the Tobacco Institute's grants to approximately 2,000 fire departments, only in the states where I was pursuing legislation. The fire service is so paramilitaristic and organized, it can get a message out within minutes. The tobacco industry knew that and put its money there.

RJR Nabisco is launching a multi-million dollar fire prevention campaign independent of the Tobacco Institute. By coincidence, it wants to start in Congressman Moakley's district in Boston: to give money to the local fire departments and to put placards, posters, little tents on tables, etc. in all of the point-of-purchase areas to show that RJR is doing its fire prevention effort.

DR. BERGMAN: Did insurance people play a role ?

MR. McGUIRE: The insurance industry was never involved. I had a horrible experience with the Fireman's Fund which was then headquartered in San Francisco. Thinking it would have an interest I asked for support on the California bill. Basically its spokesman said "No, this is a federal issue." I said, "Great, we have a federal bill," and he said, "No, because of McCaren and Ferguson, we only want to deal with it on the state level." And I said, "Well, that's a Catch-22."

MR. CARPENTER: As a representative of the insurance industry, I want to point out that there are thousands of insurance companies, as there are hundreds of different banks and other businesses. Going to one insurance company and being turned down is not the same as going to the insurance industry. You need to understand there is a package there, it's not a single outlet.

MR. McGUIRE: I think there has been a shift within the insurance industry recently, as seen in the decision to fund the Advocates for Highway and Auto Safety. And of course the work at the Insurance Institute for Highway Safety over the years is laudable. However, I have personal firsthand experience that no insurance company wanted to deal with the fire safe cigarette issue and it wasn't just the Fireman's Fund.

DR. BERGMAN: Who paid for your trips and who supported you during this long endeavor?

MR. McGUIRE: Up to 1985 there was no funding. Keep in mind, there was no formal campaign with printed brochures and the like. The work was done at places like American Burn Association meetings or a fire chiefs' conventions to which I was invited. Many of my trips were paid for that way, but most of the work actually happened over the telephone in my office. From '85 on I received some money from the MacArthur Foundation, which helped a lot.

MR. CARPENTER: When it comes to talking about political realities, I think people have to realize that in state legislatures, especially states with smaller populations, there is a reluctance to get involved in something that is viewed as a federal question. If we in the State of Washington, for example, were to require fire safe cigarettes, it might preclude sales in this jurisdiction completely. The state is not that critical to the market. California possibly can do it. We saw in the emission control area that California was able to act and lead the nation, because of the size its market.

MR. McGUIRE: I faced that question every time in every state legislature. Two classic examples were always used. The State of Minnesota set a flammability standard for tents, and it later became a national standard. And the State of Massachusetts extended the sleepwear standard for children's clothing to size 14 before it became a federal standard. Those examples always sold well in the smaller states.

# The Legislative Approach to Childhood Drowning Prevention

Garen J. Wintemute, MD, MPH

More children of preschool age drown each year than are killed as motor vehicle occupants or as pedestrians. While drowning rates are high among children in general, they are particularly high in the one to five age group. Depending on locale, their drowning rate is between four and ten times higher than for other children. In most locations, the majority of these drownings occur in a single high-risk environment: the residential swimming pool. Here in the King County area, just over half of all drownings in that age group occur in residential swimming pools. The proportion gets as high as 90 percent in Los Angeles County. The only documented exception in the continental United States is the state of Minnesota.

There is an effective environmental intervention for this problem: fencing which surrounds the pool and isolates it from the house and yard. Every community has what could be construed as a pool fencing ordinance, usually part of the county Uniform Building Code. This creates some very stiff fencing standards, but as a practical matter it refers to the backyard fence, and between 80 and 95 percent of these drownings occur to children who are in the house or in the yard, and therefore inside that perimeter fence.

## The Need for Legislative Intervention

While pool fencing is an effective preventive option, it also is costly. And cost is a substantial deterrent to the voluntary adoption of risk reduction measures. A New Zealand study, taken at a time when more than 80 percent of the population supported mandatory pool fencing, found that only 35 to 40 percent of pools had been fenced on a voluntary basis (Geddis, 1984; Horwood et al., 1981; Horwood et al., 1983). In Sacramento, Calif., only 35 percent of pool owners who favor mandatory fencing for all pools have installed a pool fence (Wintemute, 1991).

Other behavioral options, such as programs designed to increase parental vigilance, have little to offer. Numerous investigations have established that childhood drownings are associated with only brief lapses in supervision. A wide range of experience in injury prevention indicates that programs which depend on sustained enhancements in supervision do not work.

This country's legislative involvement in this area is really quite brief, making it easily the least mature story we've talked about today. However, we've benefited from the experience of several other countries, notably Australia and New Zealand, where communities with effective pool fencing laws have been compared to similar communities without those laws. It's been demonstrated pretty conclusively that pool fencing legislation will prevent anywhere from 50 to 80 percent of drownings in this high risk age group.

In this country, two states, Arizona and Indiana, have passed pool barrier codes, if you will, for residential pools. Predominantly this has been dealt with at the county level for reasons which come back to some of the general principles that Liz McLoughlin elucidated earlier. This does not imply that this is not a matter for state legislatures. It ought to be. It would save everybody a lot of time.

**Political Realities**

Brief as it is, the legislative history underscores some basic political principles. First is the importance of individuals. As you look at the counties which have passed pool fencing legislation, you can list the individuals who are responsible for the introduction and passage of that legislation. Parents and parents' groups have been absolutely vital. About five mothers of one or more children who have either drowned or suffered near drowning have taken it upon themselves not only to get legislation in the communities in which they live, but to foster repetition of that legislation elsewhere.

Not all the key individuals have been parents. A few members of the Consumer Product Safety Commission (CPSC) have committed its staff to extensive studies of the problem of pool drownings. They've determined, for example, that the estimated social cost for pool drownings in small children is between $450 and $650 million each year. They've done a careful analysis which justifies expenditure of several times the actual cost of a fence for drowning prevention measures. More importantly, they have synthesized the world's human factors literature and the experience of Australia and New Zealand and developed a model code for residential pool fencing legislation. These data were made available to us early on.

A number of us have expressed disagreement with particular elements of that code, and these same individuals in CPSC have met us halfway. They haven't changed the code, but they've gone on record as saying that it represents only a minimum. They also have disseminated our specific recommendations on how communities might go beyond the CPSC minimum. At the instigation of CPSC and some of these parents, the International Association of Building Officials adopted a very slightly modified version of CPSC's code into an appendix of the Uniform Building Code for 1991. It's not in the main body of the code because some communities may not need to adopt pool fencing regulations, but the text is there for those who want it. Many cities and counties have adopted statutes recently, and there are now a number of clearly standardized resources for jurisdictions who want to join them.

Obviously coalitions have been important too. The coalitions have been local because the activities been largely local. While the names change, the roles are the same. The local pediatric society is usually involved; the county health department is almost always involved, and in some cases a member of that department is the spearhead for the legislation. Parents are almost always active in the communities which seriously consider this legislation. Communities with lots of swimming pools tend to be affluent. That means that the parents involved tend to be well-educated, good spokespeople. Some have not only discretionary income, but discretionary time. Thus, they not only are very effective witnesses, but they are very good at mobilizing support. Elsewhere it's usually possible to find one or more senior legislators in the jurisdiction to act as a sponsor and that's obviously been helpful too.

## California Lessons

In California, the area with which I'm most familiar, there has been an attempt to form a statewide coalition. Three years ago, at our state's annual injury prevention meeting, about 100 people came to a semi-didactic session on drowning prevention. After passing around a note pad for names and addresses, we announced that the the signers had just joined the freshly created California Coalition on Drowning Prevention. It has since become the National Drowning Prevention Network. The group is now based in Texas, but remains active in California as well.

That's been both a boon and a bane for us. The coalition is made up predominantly of people with little experience in the political process. It has made a number of ill-fated attempts to draft a bill satisfactory to all members of the coalition for introduction into the legislature. That's meant, as you might imagine, the least common denominator bill with no room for bargaining. A smaller group split off from that coalition and drafted a bill which has been introduced by senior State Assemblyman Mike Roos. In retrospect I think we formed a coalition prematurely.

Our work in California also has raised the issue of how, when, and whether to involve the industry. This issue directly affects a number of consumer products industries: the pool industry, the fence industry, the pool cover industry. Despite much misgiving, the California state coalition invited a representative from the pool cover industry to sit on its board. In retrospect that was a big mistake. It resulted, among other things, in the perpetuation of the myth that pool covers are a viable alternative to pool fences as a drowning prevention measure. Most notoriously that myth has become incorporated in our state injury prevention agency's recommendations for how jurisdictions and individuals should respond to the risk posed by their own pools. Pool covers, even if motorized, require active intervention to be opened and closed. No systematic data support the contention that pool safety covers are effective at preventing drowning, while substantial experience supports the use of fencing. Numerous anecdotes, some published (Sulkes et al., 1990), have pointed to solar pool blankets, a related product, as a cause of drownings.

On the other hand, the industry inadvertently has been very helpful to us. We've been able to gain access to some industry documents, one of which definitely was not prepared for external consumption. It said, in essence, that the industry would be foolish not to go for pool fences. It pointed out, among other things, that well over half of the aggregate insurance liability associated with swimming pools was related to drownings and near-drownings among children under age five.

Another was a letter written in 1987 to our national pool industry trade association by its opposite number in Australia, which has substantially more experience pertaining to pool covers. Here is the essence of the letter from the Australasian Spa and Pool Associations to our National Spa and Pool Institute:

> The 1977 Australia Safety Cover Standard is hardly even referred to these days. We went through an era when people

believed that safety covers would in fact be the answer to child drownings in private pools, but the reality of the situation is that most children drown either during the time of the year when the pool is more often than not uncovered or at a time when the pool is actually in use by others.

Such internal documents have helped bolster our case and balance opposition by one segment of the industry against support from another. When the pool industry says, you're going to hurt our sales, we may be able to call upon the fence industry, among others, for support.

## Variations in Fencing Statutes

The provisions in fencing statutes vary widely on a number of key points, including the height of the fence. For example, the CPSC model suggests a 4-foot fence. However, empirical work (Nixon et al., 1979) demonstrates that a substantial percentage of children in the high risk age group could climb a 4-foot fence, and most jurisdictions have adopted a 4.5. foot or 5-foot requirement.

Existing laws vary in the degree to which they allow the wall of a residence to form part of a pool fence. The issue is important because the doors and windows in the wall render it more "porous" than the rest of the enclosure. It seems reasonable to allow such substitution in selected cases (e.g. when the pool is very close to the house), but eligibility should be determined by a disinterested party, such as a building inspector. Substitution should be allowed only if the incorporated part of the wall contains no doors (windows presumably are less likely to be used by children in the high risk age group). Windows must be self-closing and self-latching, and alarms must be installed.

Finally, there is disagreement as to how inclusive the fencing requirement should be. If no exemptions are allowed, there will be substantial objection from homeowners whose pools are never used by children, even as visitors. As an alternative, the requirement could extend to new pools, pre-existing pools at the time a residence is sold, and households with children five years or younger. The risk of a childhood drowning or near drowning is highest in the first months of pool ownership (Wintemute et al., 1991).

### Other Potentials for Intervention

Even among children the epidemiology of drowning varies tremendously by age group. Fencing is the one area in which a technical legislatively-mandated remedy will probably have significant impact. But among older adolescents and among adults, drowning occurs in a wide array of environments, a wide array of circumstances, and there isn't such a quick fix.

Among the possibilities is mandated CPR training for households which have swimming pools. Research done in Seattle and Sacramento, as well as some done by Consumer Product Safety Commission, documents the importance of immediate resuscitation as a major factor in determining the outcome of an immersion event (Present, 1987; Wintemute et al., 1987; Quan et al., 1990). In fact, the outcome is almost entirely determined by events which occur within the first five minutes of finding a child in the water.

In Sacramento we found that in 40 percent of the cases in which a child drowns in his own home pool, he or she is pulled from the water within five minutes of immersion, but nobody at the scene knows CPR. The vast majority - 90 percent - of pool owners believe they should know CPR, and, absent any program to encourage this, 40 percent believe that pool owners should be required to know CPR (Wintemute et al., 1990). There is a fair amount of baseline support for mandatory CPR training which could be augmented substantially. Enforcement would be problematic.

Another discrete possibility which would not be relevant in all parts of the country is guard rails. Sacramento has a river delta which forms its southwest corner, and every year the county has more drownings involving cars than involving boats. People drive off the levee roads alongside those rivers and go into the water. Most of them, about 75 percent, are drunk. There are other areas, such as Louisiana and Florida, where this is clearly a problem. On a national basis there are probably as many drownings involving cars into water as there are small children into pools. It's easy to identify the high risk spots, commonly at the place where the road takes a turn as the water course takes a turn. Guard rails at those locations could be very effective.

For other drownings, I suspect that the most profitable regulatory intervention is to focus on alcohol. Among older adolescents in the 15 to 19 age group, a quarter of all drownings are alcohol-associated. Above that, dependent on the age group, it's 40 to 70 percent. I'd recommend that we not allow the possession of alcohol at aquatic recreation sites and that we start treating boats like motor vehicles. We don't allow people to have alcohol in motor vehicles for reasons other than transport, and people haven't used boats to transport alcohol since the repeal of Prohibition. We probably ought to give some consideration to banning alcohol from water craft.

**References**

Geddis DC. The exposure of pre-school children to water hazards: the incidence of potential drowning accidents. *New Zealand Medical Journal* 1984; 97:223-226.

Hassall IB. Thirty-six consecutive under 5-year-old domestic swimming pool drownings. *Australian Paediatric Journal* 1989; 25:143-146.

Horwood LJ, Fergusson DM, Shannon FT. The safety standards of domestic swimming pools. *New Zealand Medical Journal* 1981; 94:417-419.

Horwood LJ, Fergusson DM, Shannon FT. The safety standards of domestic swimming pools 1980-1982. *New Zealand Medical Journal* 1983; 96:93-95.

Milliner N, Pearn J, Guard R. Will fenced pools save lives? *Medical Journal of Australia* 1977; 1:130-133.

Nixon JW, Pearn JH, Petrie GM. Childproof safety barriers. *Australian Paediatric Journal* 1979; 15:260-262.

Orlowski JP. It's time for pediatricians to "rally 'round the pool fence." *Pediatrics* 1989; 83:1065-1066.

Pearn JH, Nixon J. Prevention of childhood drowning incidents. *Medical Journal of Australia* 1977; 1:616-618.

Present P. Child drowning study, a report on the epidemiology of drownings in residential pools to children under age five. Washington, DC: US Consumer Product Safety Commission, 1987.

Quan L, Wentz KR, Gore EJ, Copass MK. Outcome and predictors of outcome in pediatric submersion victims receiving prehospital care in King County, Washington. *Pediatrics* 1990; 86:586-593.

Sulkes SB, van der Jagt EW. Solar pool blankets: another water hazard. *Pediatrics* 1990; 85:1114-1117.

Wintemute GJ, Drake C, Wright M. Immersion events in residential swimming pools: evidence for an experience effect. Submitted.

Wintemute GJ. Childhood drowning and near-drowning in the United States. *American Journal of Diseases in Children* 1990; *144:663-669.*

Wintemute GJ, Kraus JF, Teret SP, Wright M. Drowning in childhood and adolescence: a population-based study. *American Journal of Public Health* 1987; 77:830-832.

Wintemute GJ, Wright MA. Swimming pool owners' opinions of strategies for prevention of drowning. *Pediatrics* 1990; 85:63-69.

Wintemute GJ, Wright MA. The attitude-practice gap revisited: risk reduction beliefs and behaviors among owners of residential swimming pools. *Pediatrics* 1991 (in press).

Zamula WW. Social cost of drownings and near-drownings for submersion accidents occurring children under five in residential swimming pools. Washington DC: Directorate for Economic Analysis, Consumer Product Safety Division, 1987.

## Discussion

DR. KATCHER: We've also noted Wisconsin's lower proportion of preschool age drownings. Do pool regulations make much of an impact in parts of the country that don't publish drowning figures as much as the coastal or southern regions?

DR. WINTEMUTE: That's very possible, and we don't know because it's not been reported. The reports come from a wide range of environments though. Seattle is one. Coastal communities, not only in California, but in Texas and in Florida, report similar percentages. So do communities in warm areas which have no contact with large bodies of water whatsoever, inland California and down in the central valley. Arizona, Phoenix, and Tucson actually are highest.

At the other extreme are communities where you might expect little pool exposure because there are so many alternatives available, notably Honolulu. But there again drownings in this age group occur predominantly in residential swimming pools.

MR. MARTIN: I'm aware of one pool drowning case in the state of Washington that dealt with the lack of a fence. It was a foster home. Despite a Department of Social and Health Services regulation that required a fence around a pool before a placement would be allowed, a child was placed there and drowned. There was no fence.

DR. BERGMAN: Is the Red Cross at all interested in the drowning issue? Have you found them helpful?

DR. WINTEMUTE: Yes, to the first question; no, to the second. The Red Cross is very very much interested in the drowning issue. Just look at how many people learn to swim from the Red Cross.

Three or four years ago, it took took a serious look at its water safety programs in general. My sense from having participated in that review, is that the Red Cross relies on education and voluntary approaches. There was certainly interest in the demonstrated effectiveness of pool fencing, there was not much interest in moving from that to regulation.

DR. NELSON: The CPR point you mentioned was very good. Child and infant CPR is not part of the regular course anymore. That's going to have to be dealt with.

MR. BELLONI: There is a separate course.

DR. WINTEMUTE: That works both against us and for us in a sense. The move to take pediatric resuscitation out of CPR classes is based at least in part on the fact that CPR is a very complicated skill. It's been repeatedly demonstrated that within three months of CPR training most people who pass that training cannot perform CPR to meet the standards that they met to get their certificate. So annual retraining is probably insufficient.

On the other hand, children don't arrest the way adults with coronary heart disease do. Children, particularly in this setting, may stop breathing after immersion but continue to have a heartbeat for some minutes thereafter.

This is something that would need to be looked at a great deal further before being implemented, but it's at least an attractive possibility to consider teaching a much simpler skill, which is simply rescue breathing, to persons with pools. It would cost less and I think would be less onerous. Compliance on a voluntary basis, let alone a mandatory one, might be higher. No one has yet taken a good look at how effective that level of resuscitation should be, and that's something for the future.

# The 1988 Maryland Gun Law:
## An example of public health advocacy

Stephen Teret, J.D., M.P.H.

Probably the most critical factor involved with passage of any gun law, including Maryland's, is an incredibly strong and brutal opposition. We've heard today about the opposition of the alcohol and tobacco industries and that of motorcyclists. The gun lobby is at least equally formidable, and that's the stumbling block in passing gun laws.

**Litigation's Role**
Just as there is a synergism between federal and state legislation, there's also a synergy between litigation and legislation. That process came to play in passage of the Maryland Gun Law of 1988 (Teret, Alexander, and Bailey, 1990). It began in the early 1980s when a number of people in Maryland and elsewhere considered bringing litigation against gun manufacturers. They asked: Why should somebody be able to manufacture a weapon, particularly a weapon with very low legitimate utility, like a Saturday Night Special, without being held accountable for the results? In general Saturday Night Specials are small, short-barrelled, low-calibre guns which are easily concealed, poorly manufactured, and easily and cheaply obtained. They seem to figure disproportionately in crime.

A conference was held on this subject in Washington, D.C., in 1982, and its papers were published in 1983 in *Hamlin Law Review*. These papers suggested to plaintiffs' trial lawyers, primarily, that they bring lawsuits against gun manufacturers. One such lawsuit was brought in the state of Maryland: *Kelley v. R.G. Industries*. Kelley was a convenience store clerk who was shot in the chest by a man wielding a Saturday night special made by R.G. Industries, the largest distributor of Saturday night specials in the United States. In 1985 the Maryland Court of Appeals ruled that the manufacturer was liable (legally and financially responsible) for Kelley's injury.

That landmark decision shifted the cost of such injuries back to the manufacturer under the assumption that it was able to prevent the injuries. When that happened, R.G. Industries went out of business because they could not bear that exposure to liability.

**Legislative Response**

The judicial creation of gun control through this kind of case was extremely upsetting to the National Rifle Association and other gun enthusiasts. For the next few years, they fought the Kelley decision in the Maryland State Legislature with bills that would disallow this kind of liability. Those efforts, in turn, spurred the anti-gun forces to introduce bills limiting the proliferation of guns in the state.

In 1988, through the odd alchemy that merges bills with different goals, there was a bill introduced that would nullify the Kelley decision and create a nine-person state board to produce a roster listing those handguns which could permissibly be manufactured and/or sold in Maryland. If a particular gun was not listed on that handgun roster, it would be illegal to sell or manufacture it in the state of Maryland.

That bill became the focus of the greatest amount of legislative attention during that session. There was fighting back and forth, and the bill was amended many times on the floor. The governor came out in favor of the bill. It was touch and go until the last day of the session. On the next to the last day, Sarah Brady lent her support by visiting the Maryland Legislature, and ultimately the bill was passed and signed into law by the governor.

**Outraged Opposition**

Gun enthusiasts' outrage over the Kelley decision was nothing compared to their reaction to this law. It spawned the creation of the Maryland Committee Against the Gun Ban which vowed to overturn the law through a popular referendum. It gathered the necessary 35,000 signatures to get it placed on the ballot in November 1988 (Teret et al., 1990). It was an enormously intense campaign with media saturation. The newspapers devoted a great number of column inches to the Maryland gun law every day. By the end of the campaign the Maryland Committee Against the Gun Ban had spent $6.6 million, of which $6.3 million were contributed by the National Rifle Association. Supporters of the law raised three-quarters of a million dollars. To put that in perspective, $6.6 million was more money than ever had been spent on any campaign in the history of Maryland, including gubernatorial and Senate races.

The arguments of the Maryland Committee Against the Gun Ban were basically focused on three areas:

> First was people's need to own guns for self-protection. Television commercials conveyed this message: "Why are these people trying to take our guns away from us, the guns we need to protect ourselves?"

> The second argument, related to the first, was the alleged freedom to own guns, written into the Second Amendment of the Constitution. (I must point out that the Constitution does not give individuals the right to own whatever gun they choose. The Second Amendment has regularly been interpreted by appellate courts, including the U.S. Supreme Court, to say that in order for there to be a militia the people are allowed to have guns in the militia, but that the Second Amendment does not give people an unfettered right to own whatever gun they choose. In fact there are places, such as Morton Grove, Ill., which have said you can't own a handgun within the limits of this municipality. Both state and federal courts have upheld those laws as constitutional.)

> The Committee's third argument, which I found to be the most egregious, was based on racism. It reasoned that the law was designed (and this is true) to limit the availability of Saturday Night Specials. The handgun roster would be based on statutorily stated criteria for what guns would be acceptable for sale in Maryland. The criteria were designed to exclude Saturday Night Specials. The opponents of the law reasoned that poor people residing in the inner city (i.e. Baltimore blacks) can only afford Saturday Night Specials, so this law is therefore designed to disarm them while the more affluent suburbanites arm themselves with costlier guns. The Maryland Committee Against the Gun Ban created a newspaper for the campaign and ran headlines such as "Gun Ban Aimed at Minorities..."

Calling it a gun ban was an intentional misnomer because the law did not in any way try to ban all guns, only Saturday Night Specials. The Maryland Committee Against the Gun Ban raised the specter of a midnight knock on the door by the public health police coming to confiscate all guns. That was never the intention.

### Injury Prevention's Role

During this campaign period, the injury prevention community became galvanized to convey the notion that guns are a public health problem, not just a criminal justice problem. As the leading cause of death for some segments of the population, guns were a valid concern for the medical and public health communities. Data were presented showing the causes of death, for instance, for black males in different age groups and it was clear that guns were the number one cause of death. Our figures showed that black males were more likely to die in front of the barrel of a gun than from most other diseases combined.

Those data were impressive to the press. Our academic pie charts were converted into much more compelling graphics by the newspapers and television. The injury prevention community got an enormous amount of free time on television and radio under the auspices of presenting information of great public concern. For example, Garen Wintemute and other public health spokesmen were invited to take part in a radio talk show on the most popular station in Baltimore. The NRA was invited but didn't show up, giving us a 2 1/2 hour opportunity to discuss guns as a public health issue.

One of our reports said something to the effect that if there were a toxin in the environment, we would expect the government to dispel, clean up or eradicate it; we wouldn't tell people to be careful in dealing with the toxin or lock away the toxin in your dresser drawer and make sure that the kids don't mess with the it. Why shouldn't we do the same thing about guns if they are the vehicle of death? The newspaper ran a headline "Guns Are Toxins." It was great to have access the public so easily.

### The Results

And at the end of this long period of heavy public debate, the people of Maryland went to the voting booths (on the same election day that George Bush was elected) and the gun law was upheld by 58 percent of the voters.

Now we need to ask, what did we really win in having that law upheld? Injury prevention people never claimed that this law would produce a giant drop in the death curve. With so many guns out there already, this new law would begin to act very slowly. It only addressed Saturday Night Specials, which are but one of many types of guns and maybe not always the gun of choice for committing mayhem. Even then, gun users may have been moving toward assault weapons. While we knew it would be very difficult to show a decrease in the incidence of gun deaths, there seemed to be potential for gains that were not measurable through vital statistics. In retrospect many of us believe that has happened.

Since creation of the board, 114 types of guns have been banned from sale in Maryland. Of those, 17 were expressly denied permission to be manufactured or sold in Maryland; the other 97 were de facto banned because the manufacturers realized they wouldn't meet the criteria and never applied for permission to sell them. The 17 that are expressly denied permission include all Derringers, the Raven MP 25 and Jennings J22, which the popular press describes as guns that appear disproportionately in murders.

Probably more important is the notion now, at least within Maryland and I believe elsewhere, that the NRA is not invincible. Prior to the enactment of this gun law, the general mythology held that it's impossible to get any kind of meaningful gun legislation passed on the state or federal level. Due to the strength of the NRA, people basically gave up before they began. I'm not necessarily making a cause-and-effect relationship here, but since passage of the Maryland gun law, gun laws have been passed in other states and municipalities. There are assault weapons bans in California and New Jersey. I'm not sure I agree with them, but there are laws saying that if the owner doesn't adequately lock up a handgun and a child uses it to injure himself or another, that the owner is guilty of a crime. That law was first passed in Florida, and subsequently in other states.

I think the greatest importance of the successful passage of the Maryland law was to make people bolder. It made people think, "yes, inroads can be made". Laws can limit certain types of guns, or prohibit certain types of guns from being possessed by certain types of people. These ideas may have so much public acceptance that those laws can be passed despite the full force of the National Rifle Association and its many millions of dollars.

Punishment was a key element of the NRA strategy in Maryland. It asked its members to boycott the business contributors to our $750,000 campaign chest. When the tactic was presented in the newspaper as being unfair, the NRA said "other people boycott businesses for socially acceptable purposes, why can't we do the same?" But in the long run punishment was not successful. Witness the fact that legislators in Maryland now are willing to introduce bills for the further restriction of guns and the governor has said he will sign them.

**Preemption**

One of the issues of great importance in trying to enact state legislation regarding guns is preemption, and it's not restricted to guns. Preemption is an area of the law that can get somewhat arcane, but in its simplest explanation preemption is when a higher level of government exercises authority and therefore a lower level of government is not permitted to act. Preemption occurs on a number of different levels. For instance, the federal government may preempt the state government from doing something. This came up as an argument against fire-safe cigarettes when the tobacco companies said only the federal government could legislate cigarette safety. The argument didn't get very far, but it was used.

Another type is state versus locality, where the state tells a locality it cannot enact local legislation because only the state legislature can do that. This took place when San Francisco's handgun law was overturned by the California Supreme Court, which held that the California State Legislature had preempted localities.

The third type of preemption is legislative versus judicial. We take that example from airbag advocacy, where a suggestion once was made to sue car manufacturers for not putting airbags in cars. The car manufacturers defended themselves by saying, "No, if a court allows such a lawsuit to be won by the plaintiff, the court is, in essence, enacting motor vehicle safety standards, and the court cannot do that because Congress has preempted that field."

With guns there are all kinds of preemption arguments. At one time, state preemption laws were the top legislative priority of the National Rifle Association. It wanted to make sure that localities could not regulate local gun control, that only the state legislatures could do it. No doubt it didn't want to fight battles on so many fronts, and wanted to consolidate its battles in the state legislatures.

**The Future**

I do think that states can do a lot with gun laws. There are a lot of different types of gun laws. Some legal problems, like preemption, need to be overcome, but the bottom line is it can be done. It is being done. Whereas the NRA was once considered invincible, it now can be seen as the foe just like any other foe. A good law can be passed and even sustained by popular vote.

### References
Teret SP, Alexander GR, Bailey LA. The passage of Maryland's gun law: Data and advocacy for injury prevention. *Journal of Public Health Policy* 1990; 11(1):26-38.
Kelley v. R.G. Industries, (Md. 1985) 497 A.2d 1143.

### Discussion
DR. WINTEMUTE: I want to underscore a couple of Steve's points.

First is the degree to which there is pent-up enthusiasm in legislatures to go after this issue once it's been demonstrated that it's not suicidal to do so. The assault weapons ban in California did not pass by a wide margin. But it did pass and was signed into law by a Republican governor. Suddenly legislators came out of the woodwork to carry a whole array of bills, ranking from further legislation directed against specific subsets of guns to legislation (which passed both houses, but was vetoed by the governor) requiring that people demonstrate competency to buy a gun.

In California, we have broadened our 15-day waiting period and mandatory background check to include all guns. We also require that all gun sales occur through a licensed dealer and be reported to the state.

MS. McLOUGHLIN: That is another example of how media coverage affects legislation. The assault gun bill would never had gotten through had it not been for the Stockton school tragedy.

DR. WINTEMUTE: The bill had actually been drafted before that occurred. It's widely believed that it was a reaction to Stockton, but it wasn't. However, it's still probably the number one example of how a tragedy can help legislation. People saw that a man was able to hit 35 small moving targets in the space of about two minutes, hitting with one out of every four rounds or so.

There was further spill over to that. At the time of those shootings in Stockton, Patrick Purdy, the assailant, was legally able to buy, and was buying, guns in California. He wasn't excluded because, although he had been arrested for any number of felonies, he had never been convicted. That led to a broadening of the exclusion criteria in California.

Steve also mentioned inroads that can be made, and I want to put a sign post on one of those roads. Although there are exceptions, most of the successes (all of the successes at the federal level) in recent years have not focused on the behavior of people who use guns, but on the products themselves. Imports of assault weapons were banned by regulatory mechanism at the federal level. There was a preemptive ban (a ban before manufacture ever began) of a new type of gun which was advertised by its developer as dishwasher safe and available in an array of designer colors. That gun never came on the market. Teflon-coated ammunition, the so-called cop-killer bullet, is another example. By contrast the seven-day waiting period, the so-called Brady Bill, has been stalled for years.

It becomes important to pick issues, and some of the criteria involved in that choice are: How easily understood is this? How likely is this issue to create a schism within the so-called gun lobby? A schism has developed between the NRA and the police over a couple of these issues, which were chosen partly with that in mind.

As Steve has pointed out, the momentum is beginning to shift.

MR. TERET: I want to underscore what Garen said and share something we were told by a Washington, D.C. guru of advocacy: Sometimes it's very useful to try to produce a bill without any expectation that it will be enacted, but to try to place your opponent in an untenable position.

A great deal of value has been achieved by having the National Rifle Association oppose laws designed to eliminate cop-killer bullets. Also a great deal of value has been achieved by placing the NRA in a position of supporting plastic handguns that would be undetectable and used as terrorist's tools, or supporting guns that are going to foreseeably increase the incidence of those horribly tragic stories where a child finds a gun and thinks it's a toy and shoots himself or his playmate.

The more that you can back the opponent into a corner with a position that makes no sense or is blatantly offensive to most people's value systems, you have done a great deal for your side.

MR. FERGUSON: I think the NRA has gotten into the position of defending issues that aren't even popular among its own members. Certainly the most recent instances, where they have been at loggerheads with the police, are good examples.

We had the assault rifle bill here in Washington last year that would have passed the House easily. It failed in the Senate, but I think the public is dead set now to do something in that area. Even though the NRA has a lot of power it's not invincible.

The point about federal/state relationships is well taken. The NRA and other organizations would like to keep things at the state level as much as possible. In Washington express authority lies with the state unless it is given to the local level, and that affects a lot of the things we do. For example, a bill came up last year to allow local authorities to destroy confiscated guns, whereas by law they now are directed to sell these guns to make money. Some cities wanted the right to destroy those guns, but the bill got a hard go. It passed in the House and failed in the Senate. We're going to pass it again this year in the House, but the real issue is why not let the local authorities destroy those guns?

DR. WINTEMUTE: I feel that one of the most important things state legislatures could do now is expressly give localities the right to regulate firearms; not just the disposal of confiscated weapons, but possession, licensure and so forth. Washington is not uniform. The gun problem in Seattle is not the gun problem in Bellingham is not the gun problem in Spokane. Those cities ought to be able to treat those problems locally.

DR. BERGMAN: Does not the NRA have much more influence in rural areas?

MR. FERGUSON: In the West too, I think.

MR. CARPENTER: The perception that they are politically strong is greater in the rural areas.

MR. TERET: It's more than the perception. For instance, if you look at Maryland's referendum vote, every rural area voted down the gun law while every urban area and suburban area voted for it. But rural areas are not where the majority of gun deaths occur. Just as we have a history of local regulation for problems like restaurant inspection and rat eradication, why shouldn't we grant the same authority with regard to guns if they are one of the largest killers?

DR. BERGMAN: That is precisely the NRA's point in going to state legislatures to preempt local community options.

DR. WINTEMUTE: There's also the issue of the eroding structure of the NRA. It has lost almost a third of its membership in recent years. It's down from just over three million to just over two million, and a couple of outside polls of NRA members have shown that by roughly 60/40 margins, NRA members support the registration of all handguns, whether new or currently owned, and support the registration of all purchases of firearms.

There is a schism between the membership and the leadership. But what makes it particularly delicious is that there is a leadership fight between those now in power and the former leaders who are on the outside waiting to get back in. That group is substantially further to the right, than is the current leadership.

MR. TERET: I recently read an interview with William Batterman Ruger, the founder and chief officer of Strum, Ruger and Co., the second largest manufacturer of handguns in the United States. He was asked what he saw as the threats to gun enthusiasts and gun manufacturers. His answer (paraphrasing here) was, "that legislative threats were pretty much under control, but the real threat to the gun industry are the plaintiffs' product liability lawyers."

I think that it illustrates the point that sometimes when you're unable to accomplish public health goals through the regulatory or legislative process, it pays to use litigation. That's especially true when you have an organization as strong as the National Rifle Association with strong potential for influencing legislators. The last thing that jurors want to do is get re-elected, and they are willing to make difficult decisions. They are the conscience of the community, and by imposing costs upon gun manufacturers, they can go a long way to alleviate this public health problem.

# Concluding Remarks

Abraham B. Bergman, MD

I believe we have been successful today in focusing on the "big ticket" injury issues causing the greatest mortality and morbidity that lend themselves to approach through legislation, regulation, and/or legal action at the state level. And though public policy is not always set on the basis of facts (some would say public policy is rarely set on the basis of facts), we have learned of the utility of having reliable information at hand on the prevalence, severity, and economic impact of the particular injury problem being considered. For example we learned that data on the costs of motorcycle trauma to the public eventually turned out to be more persuasive to legislators than the ideologic issue of freedom-of-choice. The economic consequences of firearm injuries may turn out to be more persuasive than the human costs.

The need for reliable data on what constitutes effective counter-measures against drunk driving seems particularly important. It is important not to expend political capital on legislative proposals that sound good, but have not been shown to be effective. An example is the effort needed to lower the legal limit of blood alcohol from 0.10% to 0.08%, a measure which has not been shown to significantly reduce drunk driving, compared to use of sobriety checkpoints and administrative revocation of drivers licenses both of which have been shown to significantly reduce the incidence of alcohol-related trauma.

We heard much about how legislators react as human beings to real-life drama. It is imperative that injury victims be brought into the political process. The widows and children of motorcycle trauma victims at the witness table give lie to the argument that it is only the rider who suffers from his or her actions. Politicians must see individuals with slurred speech and incoordination to appreciate the devastating long-term consequences of head injuries. Politicians who help the cause, we also heard, need help at election time. Give credit loudly, when credit is due.

The importance of coalitions came up repeatedly. Since injury prevention advocates neither control large blocs of votes nor make heavy campaign contributions, we must rely, for the most part, on volunteer effort and attempting to influence public opinion. Though ultimately unsuccessful, the coalition put together in California to run the alcohol tax initiative is a fascinating example of how to broaden our bases of support.

Since we do not have dollars to purchase advertising, media news stories and editorials are virtually our only way of influencing public opinion. It helps greatly if disaster stories routinely mention whether the victim was wearing a seat belt, or a motorcycle helmet, or whether a functioning smoke alarm was present in the house, or whether the drunken driver had previous offenses. This is how the public comes to understand how lives can be saved through specific legislative actions.

Finally we were admonished that getting a law passed is invariably only a first step on the way to achieving an injury prevention goal. The law has to be enforced, and the public has to be educated before lives are saved and injuries reduced. Often the implementation efforts are more difficult than the legislative efforts.

You have been a wonderful group. I hope that the readers of these proceedings are able to gain some sense of the knowledge and enthusiasm expressed by the participants sitting around this table.

## Contributors

**Chairperson**

Abraham B. Bergman, MD

*Director*
*Department of Pediatrics*
*Harborview Medical Center*
*Professor of Pediatrics and Health Services*
*University of Washington, Seattle*

**Participants**

Nancy Bode, J.D.

*Assistant Attorney General*
*State of Minnesota*
*St. Paul, Minnesota*

Roy A. Ferguson

*State Representative*
*Bellevue, Washington*

Murray Katcher, MD, PhD

*Director*
*State Maternal and Child Health*
*Section Chief, Family and Community Health*
*Wisconsin Division of Health*
*Madison, Wisconsin*

Andrew McGuire

*Executive Director*
*Trauma Foundation*
*San Francisco General Hospital*
*San Francisco, California*

CONTRIBUTORS

Elizabeth McLoughlin, ScD
*Director*
*San Francisco Injury Center*
*Trauma Foundation*
*San Francisco General Hospital*
*San Francisco, California*

Fred Rivara, MD, MPH
*Director*
*Harborview Injury Prevention and Research Center*
*Professor of Pediatrics*
*University of Washington, Seattle*

Stephen P. Teret, JD, MPH
*Director*
*Johns Hopkins Injury Prevention Center*
*Johns Hopkins University*
*Baltimore, Maryland*

Allan F. Williams, PhD
*Vice President, Research*
*Insurance Institute for Highway Safety*
*Arlington, Virginia*

Garen Wintermute, MD, MPH
*Associate Professor*
*School of Medicine*
*University of California, Davis*
*University of California Washington Center*
*Washington DC*

**Panelists**

Scott Carpenter
*Regional Vice President*
*Insurance Information Institute*
*Seattle, Washington*

John Gullickson
*SAFECO Insurance Company*
*Seattle, Washington*

Dennis Martin, JD  *Washington Trial Lawyers Association*
*Olympia, Washington*

Susan Tracy  *Washington State Medical Association*
*Olympia, Washington*

**Observers**

James Belloni, MA  *Division of Injury Control*
David Nelson, MD  *Center for Environmental Health and Injury Control*
*Centers for Disease Control*
*Atlanta, Georgia*